GW01417603

Ship Construction Sketches and Notes

Kemp & Young

PREFACE

SHIP CONSTRUCTION SKETCHES AND NOTES was originally produced for Merchant Navy Officers who were studying for their statutory examinations and its world-wide popularity in this and many other spheres persuaded us that the content should be enlarged. However, we considered that the simplicity of the sketching and conciseness of the accompanying notes was important, and this style has been maintained. We realise that even in this enlarged book there are many alternatives to the sketches shown and it is hoped that readers will make further sketches and notes from their own observation, research and experience as and when necessary.

Since the first edition was published, metrication has been introduced and numerous changes have taken place in both Load Line and Classification Society Rules. This has necessitated a complete revision of most of the text and sketches and the opportunity has been taken to rearrange the order of presentation.

We are greatly indebted to G. W. White who has carried out the major part of the work of revision in addition to drawing all the illustrations. We are also grateful to those of our colleagues who have contributed ideas and criticism and to the many gentlemen and firms, particularly MacGregor & Co. (Naval Architects) Ltd., who have so kindly helped in one way and another.

London

September, 1971

J. F. Kemp

Peter Young

Newnes
An imprint of Butterworth-Heinemann Ltd
Linacre House, Jordan Hill, Oxford OX2 8DP

A member of the Reed Elsevier group

OXFORD LONDON BOSTON
MUNICH NEW DELHI SINGAPORE SYDNEY
TOKYO TORONTO WELLINGTON

First published by Stanford Maritime Ltd 1958
Reprinted 1972, 1973, 1976, 1978, 1981, 1983, 1984, 1987, 1989
Reprinted by Butterworth-Heinemann Ltd 1989
Reprinted 1990, 1991, 1993

ISBN 0 7506 0381 X

Printed and bound in Great Britain by
Biddles Ltd, Guildford and King's Lynn

ILLUSTRATIONS

ORDERING

A shipowner ordering a new vessel will make enquiries at various shipyards stating his requirements given in the form of a book called "Hull Specifications". This will cover the type, deadweight, speed, dimensions, draught, bunkers, holds or tanks, etc. The shipyards will then submit their estimates along with plans if these have not already been drawn up by the company's naval architect. Prior to the placing of the order, the shipowner will consider prices (and credit facilities) and delivery dates which are offered.

CLASSIFICATION

The principal maritime nations have Classification Societies whose function is to survey vessels so as to assess the adequacy of their strength and construction, and for which purpose they publish rules. The British Society is Lloyd's Register of Shipping which classes most British shipping and, as it has world-wide connections with surveyors in the principal ports, a large proportion of the world's tonnage.

The scantlings (sizes) of the materials to be used, as well as certain items of equipment (anchors, cables and warps), can be obtained from Lloyd's Rule Book, "Rules and Regulations for the Construction and Classification of Steel Ships". This is published annually with supplements between editions.

The scantlings are based on the basic dimensions of the vessel, shown opposite and defined below, detailed calculations of the still water bending moment and the section modulus of that particular item in association with other structural members.

Length L — is the distance in metres on the summer load waterline from the fore side of the stem to the after side of the rudder post or to the centre of the rudder stock if there is no rudder post. L is not to be less than 96% of extreme length on summer load waterline and need not be more than 97% of that length. Amidships is at the middle of length L measuring from the fore side of the stem.

Breadth B — is the greatest moulded breadth in metres, i.e. measured inside plating.

Depth D — is measured in metres at the middle of length L from the top of the keel to the top of the deck beam at side on the uppermost continuous deck. With a rounded gunwale D is measured to the continuation of the moulded deck line.

Draught d — is the summer draught in metres measured from the top of the keel.

A vessel built to Lloyd's highest class will be given this character,

✠ L.M.C. ✠ 100 A 1

✠ indicates "built under survey" which means that all steel was manufactured at an approved steelworks and that a surveyor supervised the building. Also all the plans will have been submitted and approved.

L.M.C. — Lloyd's Machinery Certificate.

100A — Scantlings in accordance with the Rules.

1 — Equipment in accordance with the Rules.

Vessels built for a particular type of service have a Class Notation in addition to the above characteristics, i.e. 100A 1 Liquified Gas Carrier.

In order to maintain her class a classed vessel is subject to a number of periodic surveys. These are,

(a) Annual, (b) Docking (every 12 months, maximum 24 months), (c) Special (every four years) or Continuous surveys where a maximum period of 5 years is allowed between the consecutive examination of each part. Special surveys increase in severity as the vessel gets older.

All damage must be surveyed and repaired to the satisfaction of the Society's Surveyor

A vessel which is not classed will still have to reach a minimum standard of strength and have similar surveys. Such a vessel will have a "Ship Safety Construction Certificate" issued by, or on behalf of, the government of the country of registration.

A large percentage of maritime insurance is effected at Lloyd's of London. Although the name is the same as that of the Classification Society there is no direct connection and the two must not be confused. Insurance is not even compulsory, not even "third party". Liability may be limited to 3100 gold francs (about £86) per registered ton for personal injury claims and 1000 gold francs (about £28) per registered ton for property claims.

DIMENSIONS

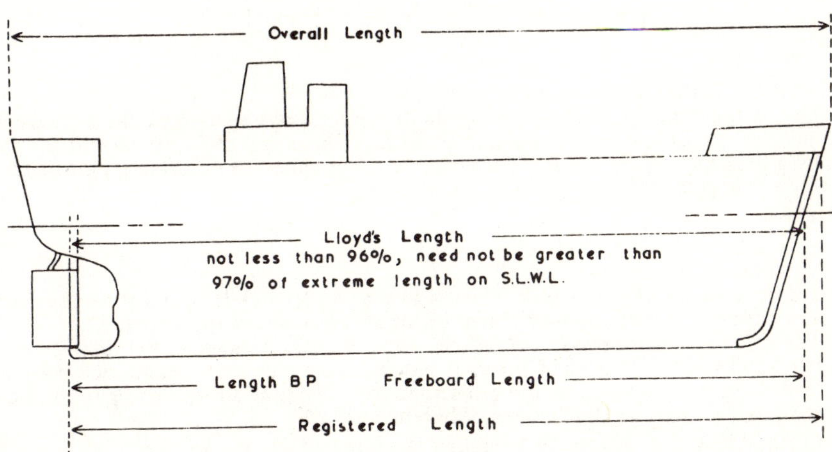

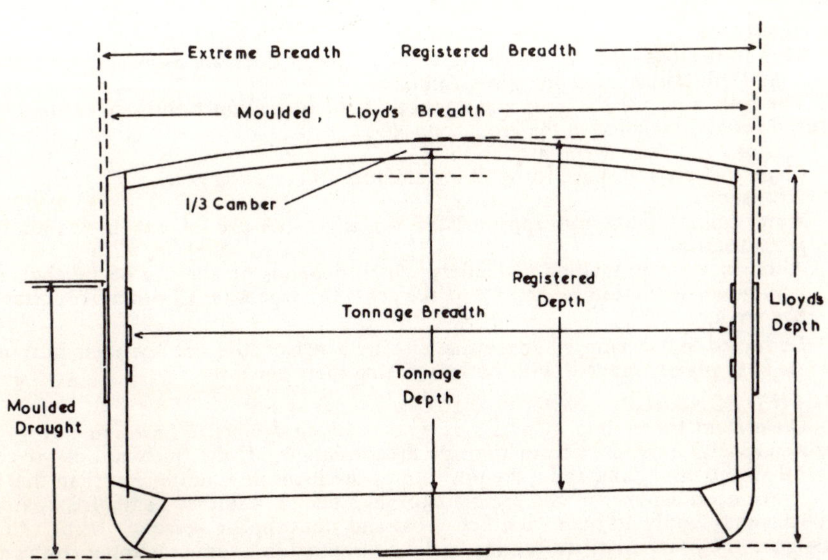

LOADLINES

Under the Loadline Rules all vessels, except Ships of War, ships solely engaged in fishing and pleasure yachts, must have a loadline. The position of this is assigned by either the Department of Trade and Industry or, in the case of a classed vessel, the Classification Society. The initial letters of the assigning authority are cut in on each side of the loadline disc.

Reference is made to the Loadline Rules and Classification Society Rules throughout the text.

GROSS AND NET TONNAGES

These are based on cubic capacity which is measured by Department of Trade and Industry surveyors and then converted to tons by dividing by 100, i.e. 100 cubic feet is equal to 1 ton. The tonnage dimensions are as illustrated on page 5.

The tonnage deck is the next deck below the upper deck; in single deck vessels this is the upper deck.

UNDERDECK TONNAGE

This is the tonnage of space below the tonnage deck bounded by the tonnage deck, the upper surface of the double bottom tanks or ceilings and the inner face of the frames or sparring plus the tonnage of shaft bossings or other appendages forming part of the hull of the ship below the tonnage deck.

GROSS TONNAGE

This is the sum of the following items :—

(1) The underdeck tonnage

(2) The tonnage of between deck space between the second deck and the upper deck.

(3) The tonnage of permanently closed in spaces on or above the upper deck.

(4) The excess of hatchways. ½% of the vessel's gross tonnage is the tonnage allowed for hatchways (other than internal hatchways which are already included) which lead to spaces included in the gross tonnage; anything above this figure is added to the gross tonnage as the "excess of hatchways".

Certain closed-in spaces on or above the upper deck are not included in the gross tonnage and these are known as Exempted Spaces. Amongst them are :— (a) dry cargo spaces, (b) wheelhouse, chartroom and radioroom, (c) galley and bakery, (d) spaces fitted with machinery or condensers, (e) washing and sanitary spaces in crew accommodation, (f) light and air spaces, and (g) water ballast tanks not appropriated for any other use.

NET TONNAGE

This is the tonnage on which port and harbour dues are assessed. It is obtained by making Deductions from the gross tonnage.

The following are the principal deductions of space which must have already been measured for and included in the gross tonnage :—

(1) Master's and crew accommodation.

(2) Chain lockers and space for working anchors and steering gear.

(3) Workshops.

(4) Water ballast tanks not appropriated for any other use (overall allowance 19% of gross tonnage).

(5) Allowance for propelling machinery which depends on the size of the engineroom; maximum deduction being 55% of the gross tonnage after all other deductions have been made.

Deducted and exempted spaces have to be a reasonable size for their purpose and are to be kept plainly marked with notices stating their purpose.

MODIFIED TONNAGE

On certain trades ship owners may find it advantageous to have some of the ship's carrying capacity exempted from tonnage measurement. If the freeboard of the vessel is greater than the minimum, and if the position of the loadline is not higher than that which would have been assigned if the second deck had been considered as the freeboard deck, the owner may apply to have modified gross and net tonnage assigned. If such tonnages are assigned the gross modified tonnage will be obtained in a similar manner to the normal gross tonnage, except that item 2 will be omitted and whenever reference is made to the upper deck the second deck will be substituted. Deductions to obtain the modified net tonnage will be similar to those for the normal net tonnage.

TONNAGE MARK

This mark must be cut in on each side of the vessel when modified or alternative tonnages have been assigned. If the vessel has a modified tonnage the mark will be cut in on the level of the deepest draught to which she can load. If alternative tonnages are assigned the distance of the mark below the tonnage deck line is found from tables.

ALTERNATIVE TONNAGE

A ship having loadlines in the normal position, (i.e. the freeboard deck being the upper deck and the vessel of maximum strength) with minimum freeboard may, on the application of the owner, be assigned an alternative. This tonnage is calculated in the same way as the modified tonnage detailed above. Where assigned, the alternative (smaller) tonnage is applicable if the tonnage mark is not submerged whilst the normal tonnage applies when the tonnage mark is submerged.

SUEZ CANAL AND PANAMA CANAL TONNAGES

These differ in certain respects from British Tonnage and the appropriate Tonnage Rules must be studied to obtain full details.

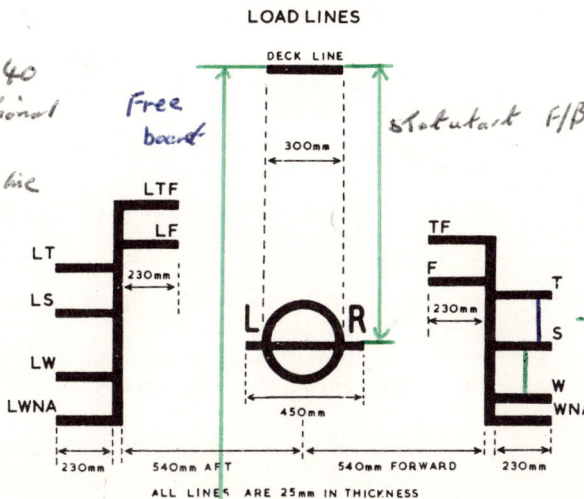

LOAD LINES

DECK LINE

ALL LINES ARE 25mm IN THICKNESS

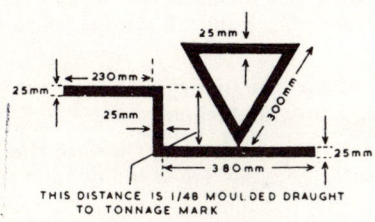

THIS DISTANCE IS 1/48 MOULDED DRAUGHT TO TONNAGE MARK

TONNAGE MARK

NEW TONNAGE RULES

I.M.C.O. held a Conference in June 1969 to establish a universal system of tonnage measurement for ships engaged on international voyages.

The Conference accepted three Recommendations; they are :–

(1) Acceptance of the International Convention,
(2) Uses of Gross and Net Tonnages,
(3) Uniform interpretation of definition of terms.

GROSS TONNAGE means the measure of the overall size of a ship. It will be obtained from a formula based on the volume of all enclosed spaces in the ship and a constant which is to be either calculated or tabulated.

NET TONNAGE means the measure of the useful capacity of the ship and will be obtained from a formula. This formula is based on the volume of cargo spaces in cubic metres, moulded depth and draught, number of passengers if over 12, and certain constants.

The Convention, when ratified, will apply to new ships, existing ships which undergo alterations and all existing ships 12 years after the date on which the Convention comes into force.

LIGHT DISPLACEMENT is the weight of the hull, engines, spare parts, and with water in the boilers and condensers to working level.

LOAD DISPLACEMENT is the weight of the hull and everything on board when floating at the designed summer draught.

DEADWEIGHT CARRYING CAPACITY is the difference between the light and loaded displacements and is the weight of cargo, stores, ballast, fresh water, fuel oil, crew, passengers and effects on board.

STATUTORY FREEBOARD is the distance from the upper edge of the summer load line to the upper edge of the deck line.

RESERVE BUOYANCY is virtually the watertight volume above the waterline. For fuller details with reference to the assignment of loadlines the reader is referred to the Loadline Rules.

SHEER may be defined as the rise of a vessel's deck fore and aft. It adds buoyancy to the ends where it is most needed. A correction for sheer is applied when calculating the freeboard.

CAMBER OR ROUND OF BEAM helps to shed water from decks and to strengthen the upper deck, upper flange of ship girder, against longitudinal bending stresses, especially compression.

FLARE or FLAM is the 'fall out' of the vessel's bow plating. It increases buoyancy thus helping to prevent the bow from diving deeply into head seas. It also increases the breadth of the forecastle head and allows the anchors to drop clear of the shell plating.

TUMBLEHOME is the 'fall in' of the side plating from the waterline to the upper deck. Modern vessels have little, if any, tumblehome.

RISE OF FLOOR is the distance from the 'line of floor' to the horizontal, measured at the ship's side. The object is to allow liquid in the double bottoms to drain to the centre line.

The remaining items, which are illustrated, are used in design and are self explanatory.

TERMS

AFTER SHEER
FORWARD SHEER
RAKE
CUT UP
CUT UP

FLARE

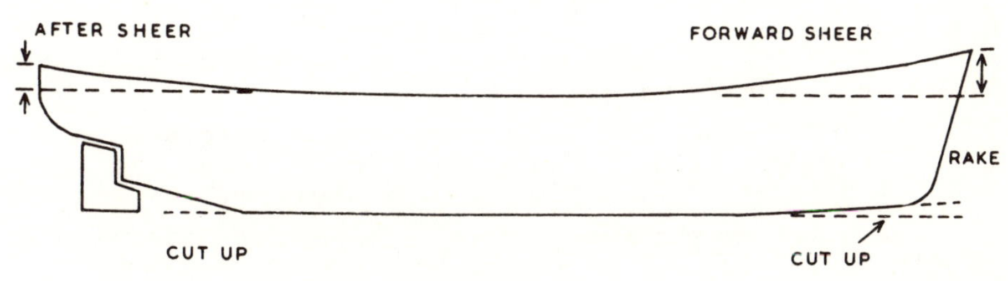

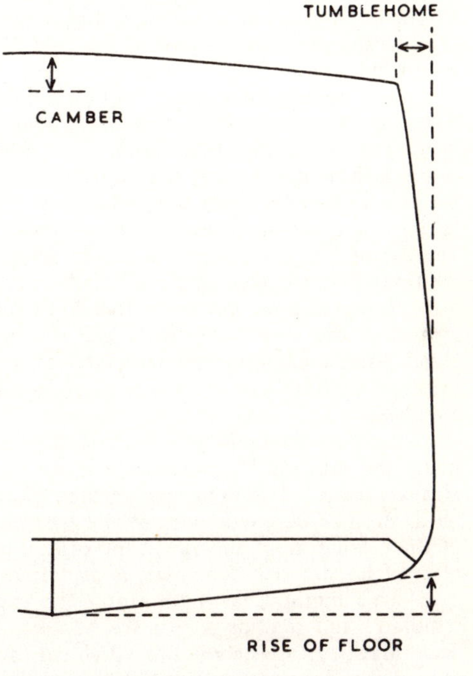

TUMBLEHOME
CAMBER
RISE OF FLOOR

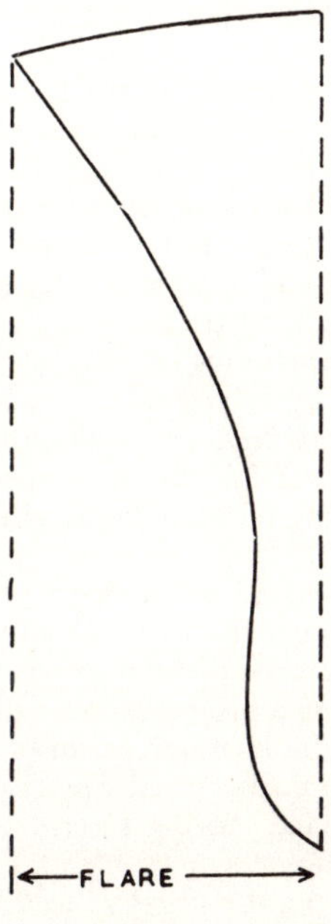

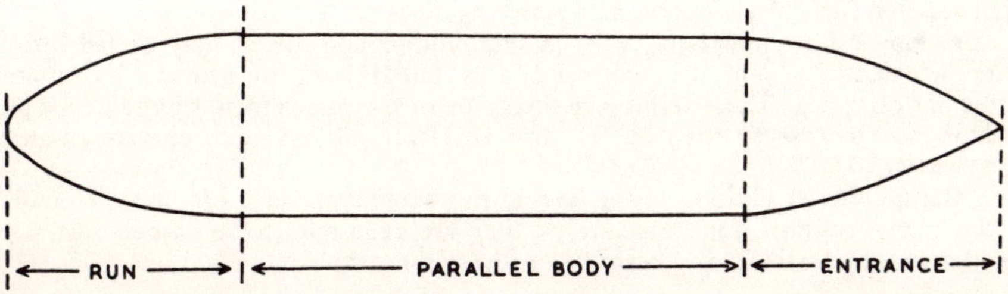

RUN
PARALLEL BODY
ENTRANCE

SHIPYARD PRACTICE

The shipyard will start drawing the final plans for approval by the owners after the order has been placed. Lloyd's Register will require to approve the plans if the vessel is to be built under survey.

The form of the ship is delineated on a scale drawing known as a Lines Plan. A set of Lines consists of three views as follows :-

Profile, or Sheer	–	side elevation (starboard)
Half Breadth Plan	–	plan view
Body Plan	–	cross sectional view

The finished lines plan must be fair i.e. all the curved lines must run evenly and smoothly and there must be exact agreement between corresponding dimensions of the same point shown in the different views.

This is done by loftsmen who lay the lines down on the floor of the Mould Loft from offsets given by the drawing office. The Mould Loft must have a large area free from obstructions and it is often over the plating shop. The floor consists of close fitting tongued and grooved boards laid diagonally over which are laid large squares of plywood or a second layer of tongued and grooved boards. The upper section forms the Scrieve Boards which will be dismantled later and set up alongside the bending slabs.

The vessel's lines are set out on the loft floor in a similar manner to the Lines Plan, but using frames instead of sections as a measure of the longitudinal position of any point in the vessel's length. It is usual to combine the Half Breadth Plan and Sheer Plan on a common base line so that they are superimposed. Because of space restrictions the Half Breadth and Sheer Plans are frequently reduced in size, i.e. to a quarter full size. The fairing of the lines then takes place and once completed the Body Plan is set out separately. The Frame Lines are then run in and scribed on the Body Plan. After the completion of the Body Plan a corrected set of offsets is then returned to the drawing office.

Templates are made for the ship's plates and framing by reference to the lines as scribed in the mould loft floor and the various working drawings, i.e. shell expansion plan. After completion the templates are sent to the appropriate fabrication shops where they are used as patterns for the cutting, shaping and bending of the various structural members.

Modern methods now include the use of an optical system of marking off. Initially the ship and all its parts are drawn one-tenth full size in pencil or ink, often on to a plastic surface. The drawings are then photographed on to a glass plate negative at one-tenth the size of the drawings. The glass plate is then placed in a projector, near the roof of the plating shop, and the shape of that piece of the structure is projected on to a steel plate of suitable size. The plate is then marked off and is then ready for cutting.

This method saves time and the expense of making and storing scrieve boards, templates and moulds.

Machines are now in use which cut the plates directly from the negative or drawing. Movement of the cutting head and hence the shape of the plates is governed automatically by means of a sensitive photo-electric head which follows the required outline shown to small scale on films, drawings or glass plate negatives.

Computers are now being used in shipbuilding and offsets may be fed into a computer which then fairs the lines and uses faired lines to provide information, in mathematical terms. These terms give the shape of the plate etc. in symbols on a punched card or tape which can then be fed into an electronic device to operate an automatic flame cutter.

Computers, in addition to the lines fairing programme etc., can also materially assist in the many routine ship calculations that are required to be carried out, i.e. stress calculations, hydrostatic and stability information etc.

LINES PLAN

SHEER PLAN

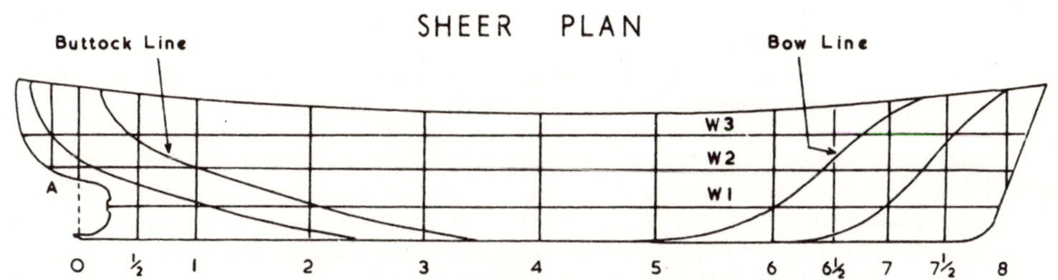

HALF BREADTH PLAN

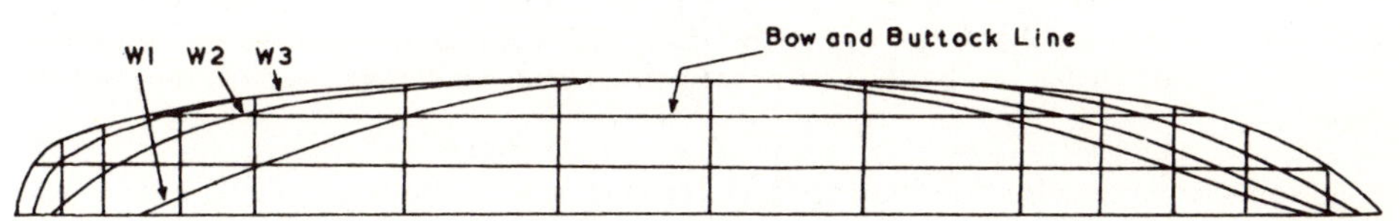

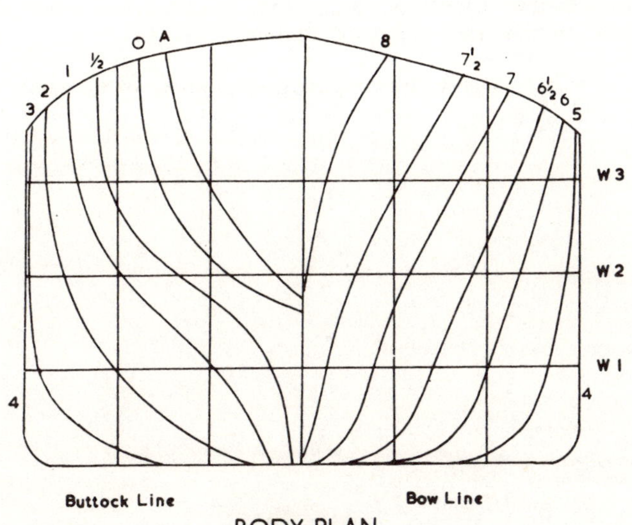

BODY PLAN

Plates are nearly always shaped cold, large machines being used for this purpose. Frames are also usually turned cold and may be cut out of a flat plate. Transverse framing requires more turning than longitudinal framing.

Should a frame have to be turned hot it will be heated in the furnace and then brought out onto the bending slab where a set bar of the correct shape, as obtained from the scrieve board, will have been pinned. The frame is then "pushed" by the squeezers until it is touching the set bar along its length. This takes some time and as each section is turned it is secured in position by dogs. Reheating may be necessary.

Modern methods of welded construction have radically changed shipyard procedure and the important consideration is to ensure that the correct pieces arrive at the welding shop at the correct time. After welding the prefabricated sections, some up to 800 tonnes in weight, will be taken to the berth for erection. Usually centre sections are erected first, then work proceeds fore and aft and sideways.

When the vessel is nearly ready for launching, launching ways will be set up. These consist of a fixed portion on the ground and a sliding portion attached to the ship, between the two portions is a layer of thick grease. Shortly before the launch the vessel is transferred from the building ways to the launching ways and from these she is put into the water. At the fitting out basin further specialised trades will take over to complete the vessel.

Some larger vessels are built in drydocks instead of being built on slipways. It is then only a matter of flooding the dock to carry out the "launching".

Where building berths are inadequate to cope with the dimensions of modern large vessels, especially tankers, it is not uncommon to build the hull in two sections. The two parts will be launched, floated together, carefully trimmed and then welded together. The two part building method has the advantage that launching half a ship of very large size is an easier operation than when launching a full length vessel.

The above is only a very brief description of some of the work undertaken in a shipyard.

FRAME BENDING

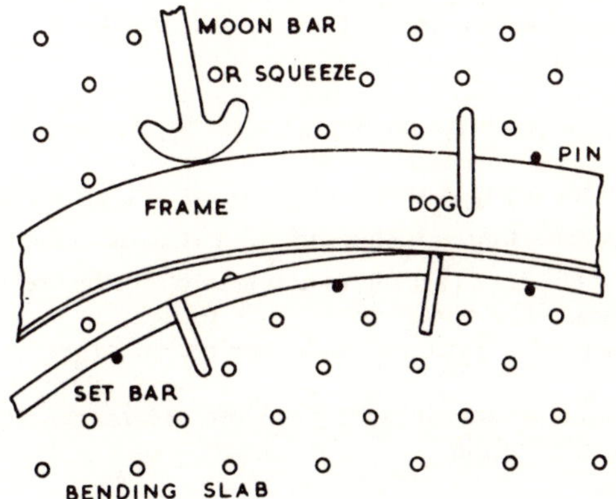

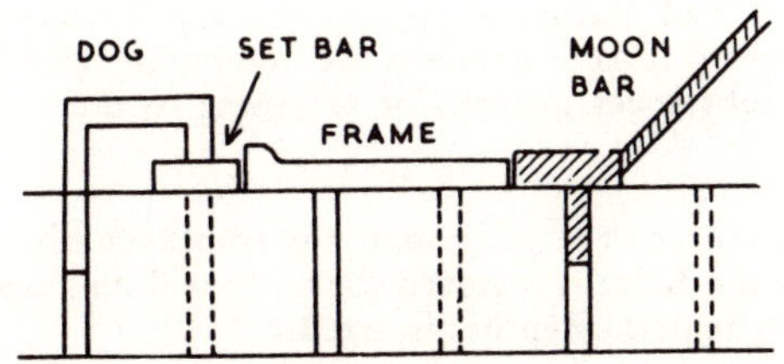

LAUNCHING WAYS

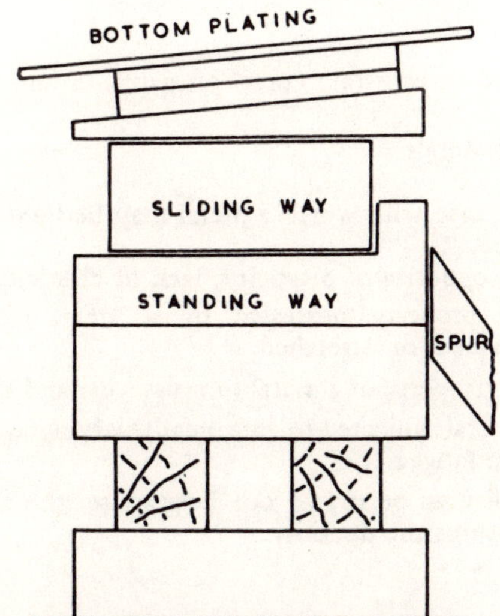

STRESS

Stress	is load or force acting per unit area and is usually (Lloyd's Rules) expressed in kilogrammes per square millimetre, kg/mm^2.
Strain	is the distortion in a material due to stress. Stresses are of three main types,

(1)	Tensile	forces acting in such a direction as to increase the length.
(2)	Compressive	forces acting in such a direction as to decrease the length.
(3)	Shear	the effect of two forces acting in opposite directions and along parallel lines. The forces act in such a direction so as to cause the various parts of a section to slide one on the other.

Stress is proportional to the distance from the neutral axis; the neutral axis passes through the centroid of the section.

Strength of materials.

When a force, or a load, is applied to a solid body it tends to change the shape of the body. When the applied force is removed the body will regain its original shape. This property, which most substances possess, of returning to their original shape is termed 'elasticity'.

Should the applied force be large enough, the resistance offered by the material will be overcome and when the force is removed the body will no longer return to its original shape and will have become permanently distorted.

The point at which a body ceases to be elastic and becomes permanently distorted is termed the 'yield point' and the load which is applied to cause this is the 'yield point load'. The body is then said to have undergone 'plastic deformation or flow'.

Whenever a change of dimensions of a body occurs a state of strain is set up in that body.

See illustrations and 'stress-strain curve' on pages 15 and 17.

Mechanical properties of Metals.

Plasticity	The ease with which a metal may be bent or moulded into a given shape.
Brittleness	The opposite of plasticity, lack of elasticity.
Malleability	The property possessed by a metal of becoming permanently flattened or stretched.
Hardness	The property of a metal to resist wear and abrasion.
Fatigue	A metal subjected to continually varying loads may eventually suffer from fatigue.
Ductility	Ability to be drawn out lengthwise, the amount of the extension measures the ductility.

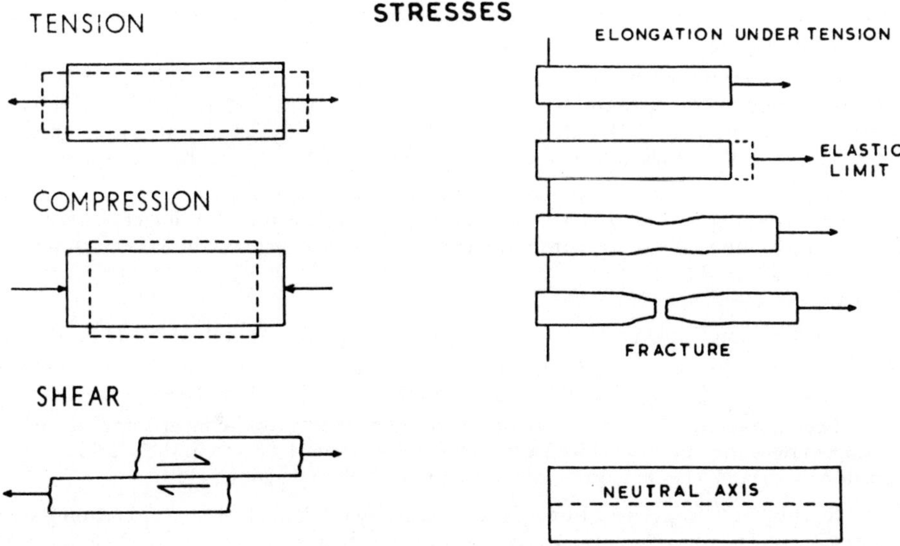
STRESSES
TENSION
COMPRESSION
SHEAR
ELONGATION UNDER TENSION
ELASTIC
LIMIT
FRACTURE
NEUTRAL AXIS

Brittle Fracture.

When a tensile test is applied to a material it elongates elastically, then plastically and finally it fractures. A warning of fracture is given by elongation.

Occasionally mild steel behaves in a completely brittle manner. The fracture occurs without warning when the load is well below the elastic limit of the metal. A fracture of this type is known as 'brittle fracture'.

Brittle fracture occurs in both welded and riveted construction. Since welded plates are continuous brittle fracture may travel through the plates with disastrous results. In riveted construction the crack only travels to the edge of the plate in which it commenced. Once a crack starts is may travel at a very high speed, up to 2000 metres per second.

Significant factors for the occurrence of brittle fracture are (a) low temperatures, at or near freezing point, (b) load on material relatively light, (c) defects or faults in a weld, and (d) internal stresses within the material, i.e. welding may initiate a fracture.

STEEL

Steel is manufactured by the purification of pig iron. It is to be made by the open hearth, electric furnace, oxygen process or other approved processes. The manufacturing process is to be such as to minimise the non-metallic content of the steel.

The characteristic of steel may be varied by changing its chemical composition; the tensile strength may be altered by varying the percentage of carbon in the steel and by the addition of various elements such as chromium, nickel, manganese, etc.

The physical properties of steel depend basically on the percentage of carbon present, increased carbon content, increased hardness of the steel.

In shipbuilding mild steel is generally used. It is relatively cheap, may be rolled, forged and welded, and worked cold or hot without any appreciable loss in its mechanical properties. At low temperatures mild steel lacks notch toughness and is subject to brittle fracture.

There are five grades of steel, A to E, used in shipbuilding, the grades varying according to the chemical composition. Grades A and B are ordinary mild steel, grades C, D and E possess higher notch toughness characteristics.

The Rules state the particular grade to be used, where greater strength is required a higher grade is used. Vessels over 200m in length are to have grade E steel used for the sheerstrake, bilge strake and keel for a distance of 0.4L amidships.

In large oil tankers, ore carriers, etc., high tensile steels are used. These are special steels which in addition to having an increased strength over mild steel retain this strength at low temperatures. The use of high tensile steel in large vessels permits of a reduction in plate thickness and hence a saving in weight. The high tensile steel used in ship construction is capable of being fabricated and welded under shipyard conditions.

All steel is subjected to tests, these being tensile, (41–50 kg/mm^2, elongation 22%), bend and impact tests. The impact test is only carried out on the higher grades of steel and the temperature at which the test is to be carried out is given i.e. Grade E steel, -10°C.

Steel is subjected to heat treatment depending on the grade, i.e. normalizing or annealing. The purpose is to produce a steel having a fine grained structure; it improves its tensile strength, ductility and resistance to shock.

Annealing is the uniform heating of the material and then allowing it to cool slowly.

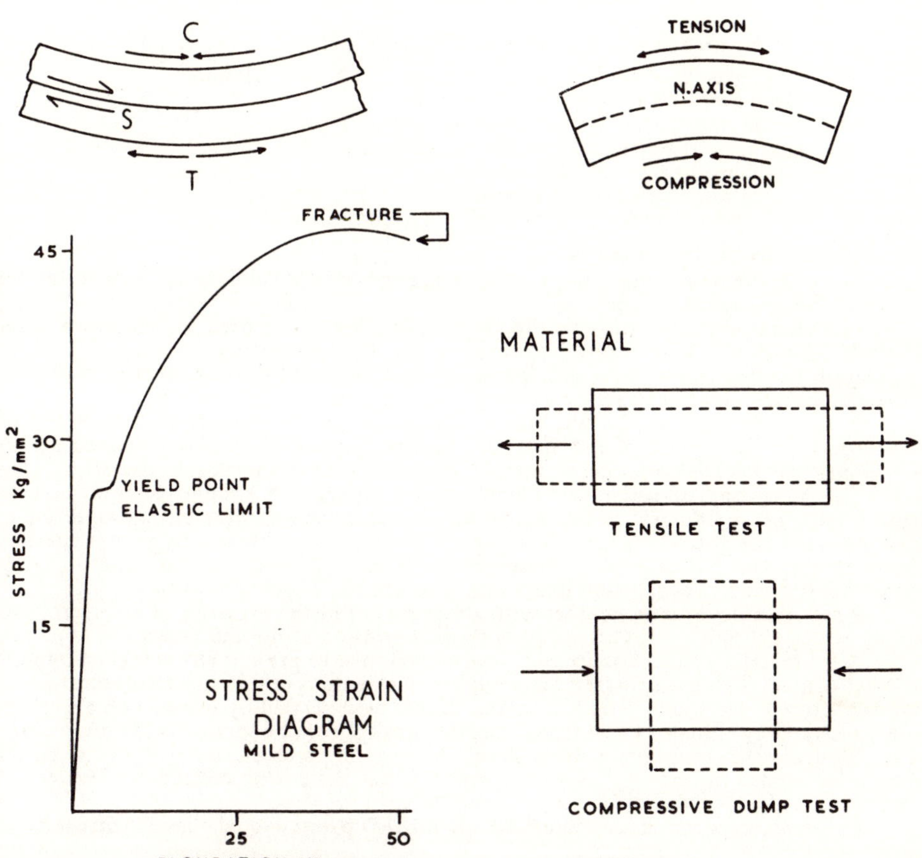
STRESSES
C
S
T
TENSION
N.AXIS
COMPRESSION
FRACTURE
45
30
15
STRESS Kg/mm²
YIELD POINT
ELASTIC LIMIT
STRESS STRAIN
DIAGRAM
MILD STEEL
25
50
ELONGATION IN mm
MATERIAL
TENSILE TEST
COMPRESSIVE DUMP TEST

PLASTICS

There is a wide and varied use of plastics in shipbuilding. The many different plastics in use have many different properties. Basically they are of light weight, flexible, durable, not highly inflammable, have good insulating properties, and give ease of fabrication.

A few of the uses to which plastics may be used on board are given below :–

fibre glass used for the construction of lifeboats, insulation
laminated plastic bearings – stern tube
nylon, terylene, polypropylene etc. – mooring ropes
piping – non-essential services
cable insulation – electric wiring
insulation – refrigerated compartments
decorative laminates and fittings – accommodation.

ALUMINIUM IN SHIPBUILDING

Most of the aluminium alloys used at present in shipbuilding are mainly for the construction of superstructures.

The chemical composition of the aluminium alloys to be used is laid down in the Rules along with the various tests that sample pieces have to undergo.

Both riveting and welding may be used in the joining of the various structural parts though in present day construction welding has mainly superseded riveting.

A major problem in the welding of aluminium is the removal of the oxide film that forms on the surface of the material. Satisfactory welds are now obtained by the use of inert-gas shielded arc process (metal inert-gas or tungsten inert-gas process).

The advantages of using aluminium alloy, especially for superstructures, are its light weight, i.e. a reduction in top weight and therefore an improvement in stability plus increase in deadweight, resistance to bending stresses and its corrosion resistance properties. Disadvantages are that it is more expensive to work and weld, different coefficient of expansion between steel and aluminium may give rise to a hogging condition.

When aluminium is in contact with other metals in the presence of an electrolyte, i.e. sea water, galvanic corrosion occurs with the corrosion of the aluminium.

All bi-metallic connections should be insulated so as to prevent any actual aluminium to steel contact. This is of paramount importance at the sides of superstructures which are subjected to sea water. The two metals should be separated by using a non-absorbent joint, i.e.Neoprene sheet. In addition the steel work should be sprayed with aluminium, zinc or galvanised. Aluminium rivets should be used in preference to steel rivets and the aluminium lapped on the steel in such a manner that there is no collection of water, i.e. alumium fitted on weather side.

In the preservation of aluminium LEAD BASED paints should NEVER be used.

Measurement of Sectional Strength.

A beam subjected to a load tends to bend, the extent to which this occurs depends on the resisting moment. The resistance to bending is a function of the material from which the beam is constructed and its geometrical property. The factor which relates to its geometrical form is termed the 'moment of inertia' of the beam.

The moment of inertia "I" is a measure of a beam's ability to resist deflection; it is an indication of how the mass is distributed with respect to the neutral axis. With a given cross sectional area it is possible to create a number of different sections. One section will have a greater I than another because of the greater distances of its flanges from the neutral axis.

The distance of the upper (or lower) flange from the neutral axis (designated by "y"), is an indication of the efficiency with which the flange can resist stresses due to bending.

If the moment of inertia I is divided by y the resultant expression I/y can be used as a standard or modulus of the ability of a section to withstand bending and associated stresses. The expression I/y is termed the section modulus.

Although the geometrical distribution of material in a section is a measure of the strength of the section, the material used in the section also determines the strength of the beam. The greater the strength of the material the greater will be its resistance to bending.

Resistance to bending implies stress (p), the maximum stress occuring at the uppermost and lowermost parts of a loaded beam.

$$\text{Total resistance to bending} = \frac{I}{y} \times p$$

$$\text{Moment tending to bend beam} = M$$

$$\text{then} \quad M = \frac{I}{y} \times p$$

If the bending moment and maximum stress which can be permitted (due to material) is known,

$$\text{then} \quad \frac{M}{p} = \frac{I}{y} = \text{section modulus}$$

It is then necessary to select a section with a Section Modulus at least equal to that required i.e. reference to 'geometric properties of Rolled Sections and Built Girders'.

In obtaining the scantlings of the various structural items the section modulus of that particular section in association with other structural members has to be calculated.

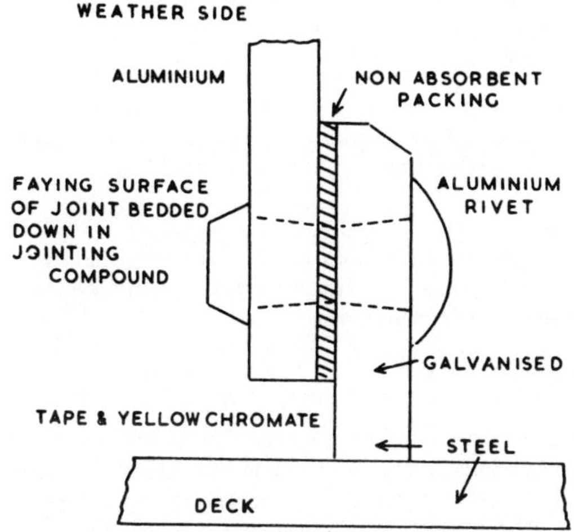

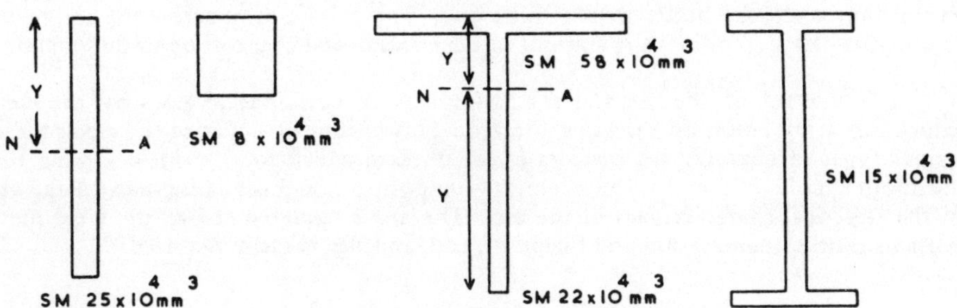

WELDING

This is the fusion of two metals together by the application of heat resulting in a joint which is as strong, or stronger, than the metals being joined. Welded joints may be obtained with any metal although some are more difficult to weld than others.

The heat required to produce the weld may be generated in a number of ways, examples are :—

(a) Electric Arc Welding.
This is the standard method used in shipbuilding and repair work, detailed by the Classification Society and is the process described below.

(b) Gas Heating by Oxygen and Acetylene.
This method has now been superseded by electric arc welding in shipyards, the oxy-acetylene blowpipe or torch being mainly used in the cutting of steel.

(c) Resistance Welding.
In this method the parts to be joined are clamped together and an electric current (AC) is passed through the joint. The resistance to the passage of current across the joint creates heat thus causing the metal to melt with resulting fusion. Spot welding is a similar form of resistance welding.

(d) Thermit Welding.
A combination of chemicals called the thermit is fired producing a chemical reaction. It is essentially a casting process and this method is mainly used in the joining of castings, i.e. sternframes.

Electric Arc Welding.

An electric arc is formed when an electric current passes between two electrodes separated by a short distance from each other. In electric arc welding one electrode is the welding rod while the other is the metal (plate etc.,) to be welded. The electrode and plate are connected to the supply, an arc is started by momentarily touching the electrode on to the plate and then withdrawing it about 3 to 6 mm from the plate. A spark forms across the gap, the air surrounding the spark becomes ionised and current flows across the gap. The temperature created is approximately 4000°C, the current flow between 20 to 600 amperes depending on the thickness, type of metal being welded and voltage, AC or DC. If the current is too high a thick layer of weld metal will be deposited and spatter may occur, i.e. small particles of metal scattered around the weld. In general several thin layers are better than a thick one. The voltage drop across the arc, between 15 and 40 volts, determines the amount of penetration and shape of deposited metal.

The control of the arc and the absorption of atmospheric gases by the weld is reduced to a minimum by shielding the arc. This is done by covering the electrode with various types of coatings, i.e. inert gases are released which form a shield around the arc and molten pool of metal. The chemical composition of this coating has a large effect on the electrical characteristics of the arc. The arc is rendered stable, the weld metal is easier to control, penetration and fusion is good, and slag is easily removed.

ELECTRIC ARC WELDING

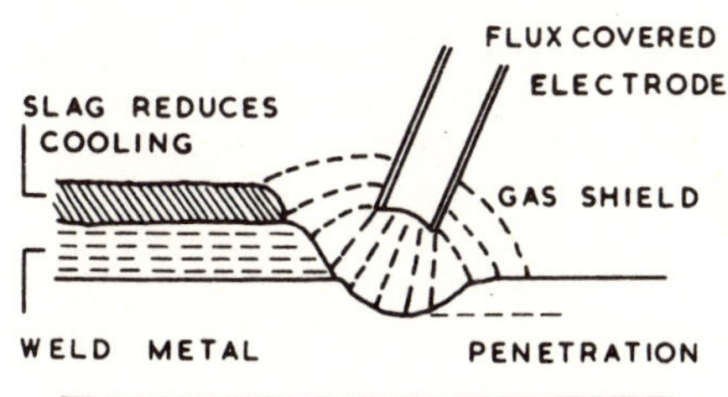

GOOD WELD

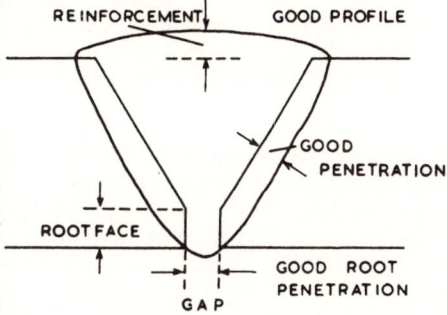

REINFORCEMENT
GOOD PROFILE
GOOD
PENETRATION
ROOT FACE
GOOD ROOT
PENETRATION
GAP

FAULTS

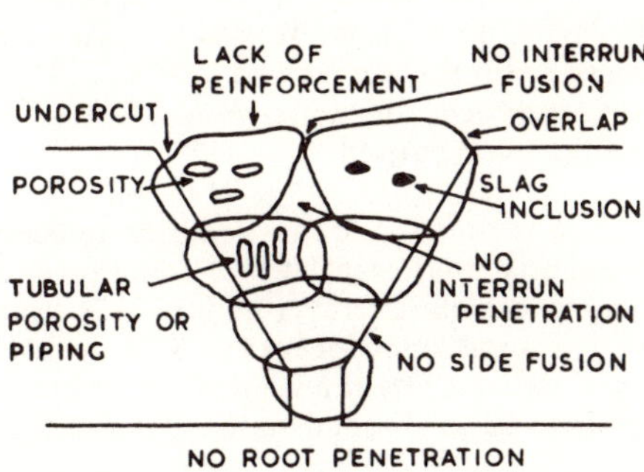

PLATE EDGE PREPARATION

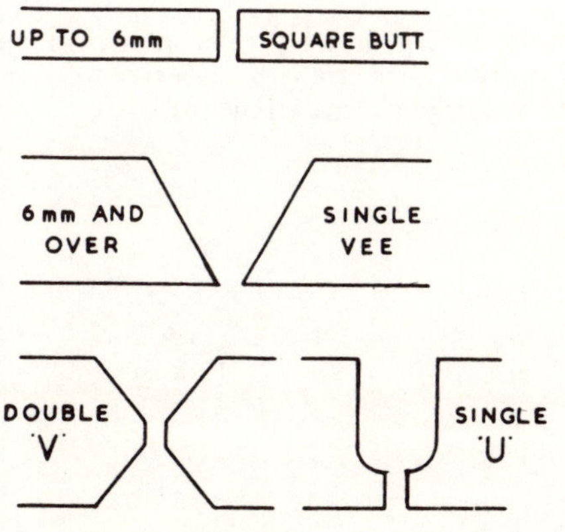

TRIPLE HEADED BURNER
FOR QUICK EDGE PREPARATION

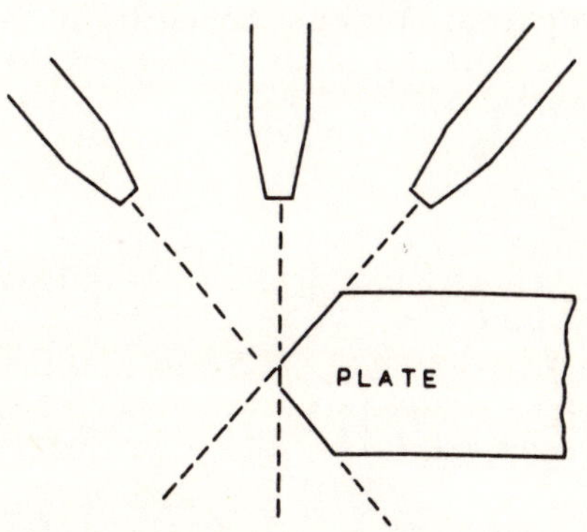

WELDING POSITIONS

Downhand welding is easier to carry out and the various parts are placed in the most convenient position when fabricating. For instance, the bottom of a ship will probably be welded in the upside down position, the section then being turned over for removal to the berth for assembly.

Vertical welding may be performed either upwards or downwards. Upwards welding is most common; as the weld metal is deposited it is used as a step on which to continue the deposit.

Overhead welding is a difficult operation to perform. Lightly coated rods, correct current control and a short arc are essential for a satisfactory weld.

Machine welding is now extensively used, excellent results being obtained providing the machine is correctly set up. The electrode is wound on a drum in a continuous length. In the submerged (union melt) process the arc is submerged under a fusing powder which runs down from a hopper on the machine. The unfused powder may be swept up and re-used.

The number of runs required to complete the weld will depend on the electrode. It is essential that each layer is completely deslagged and brushed before making the next deposit otherwise imperfections will occur in the resultant weld. On completion the weld is sometimes 'peened'. This consists of lightly hammering the weld and adjacent metal either when the weld is hot or immediately it has cooled in order to relieve stresses present and to consolidate the metal's structure.

The materials to be joined must be properly prepared, plate edge preparations are illustrated.

The quality of a weld is dependent upon the training and efficiency of the operator. A good weld is stronger than riveting but it is difficult to test. The inspection takes the form of a visual examination supplemented in the case of important structural welds by non-destructive tests or methods of flaw detection. These may consist of radio-graphic examination (X-rays), magnaflux (magnetic method), etc.

The advantages of welding are a substantial saving in weight, economy in the construction, greater fabrication potential, better continuity of strength, a greater freedom in design and the ease of obtaining an oiltight and watertight connection.

SECTIONS USED IN WELDED CONSTRUCTION

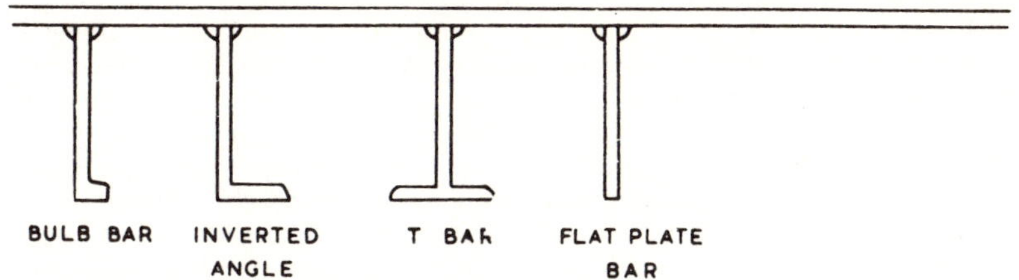

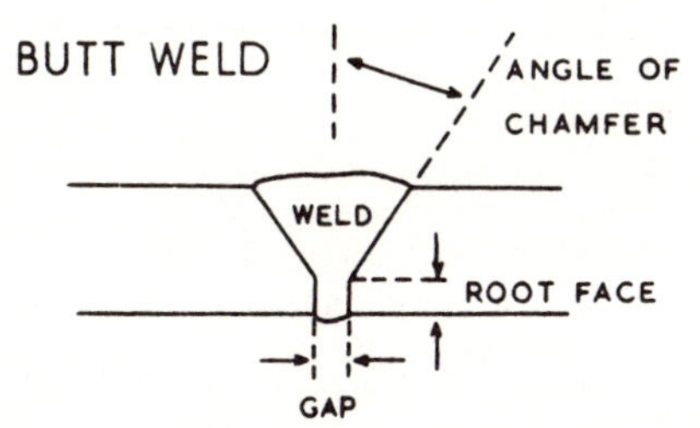

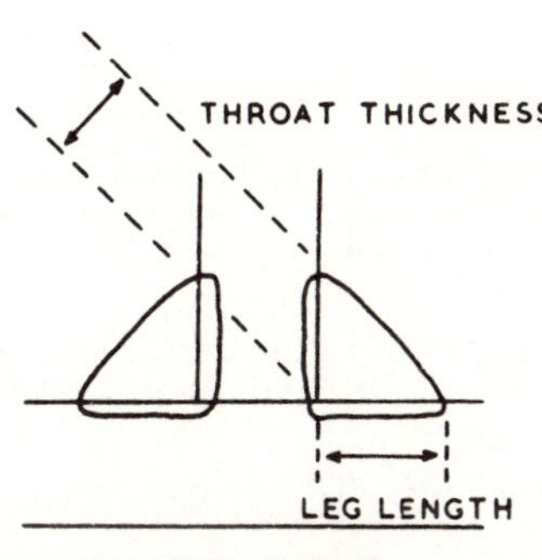

FILLET WELD

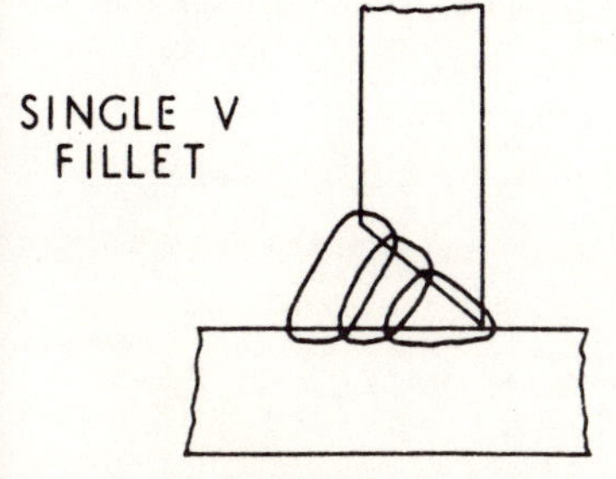

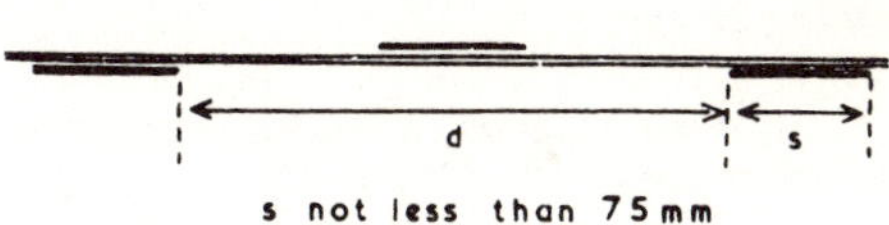

s not less than 75 mm

CHAIN INTERMITTENT WELDING

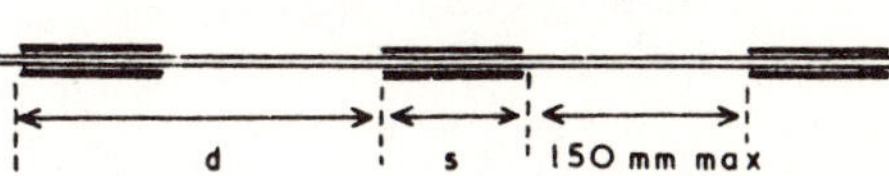

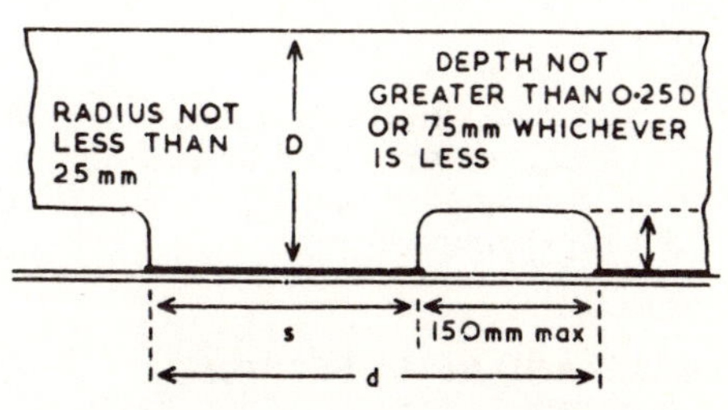

SCALLOPED FRAMES, LONGITUDINALS WITH DOUBLE FILLET WELDS

d given in Rules

RESISTANCE WELDING OF A SEAM

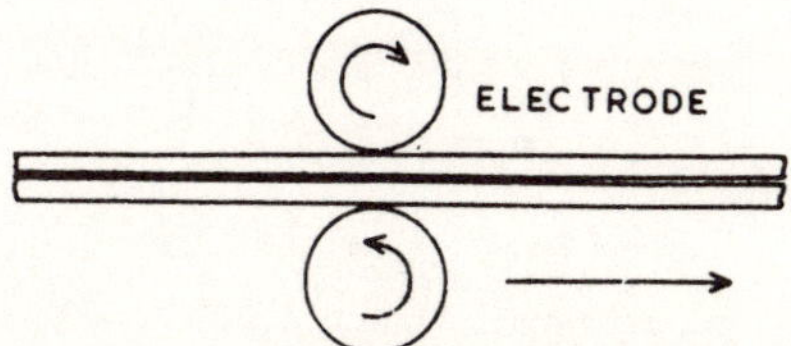

Disadvantages are due to locked up stresses occuring on the cooling of the weld and distortion and buckling of plates due to the rapid heating and cooling of the weld. It is essential to avoid design notches and hard spots and special attention must be given to the welding sequence.

Design Notch: any structural discontinuity or any abrupt change in the geometry of the section may constitute a design notch. Example: square cut corners of hatchways used in rivet design are too severe for welded construction, the corners of all openings should be well rounded. See illustrations.

Hard Spot. regions of stress concentration found in welded construction, i.e. reduce the strength of bracket or stiffener gradually by shaping, see illustration.

Welding Sequence.

In order to avoid distortion the amount of welding should be kept to a minimum, i.e. use of larger size plates. The welding sequence should be simple and practical, weld the butts before the seams so that the plates or panels are free to move and there is no restraint during welding operation. In the case of frames or longitudinals and shell seams, the frames may be scalloped in way of the shell seams. Twisting may be prevented by making all the welds in one direction.

Repairs.

The general principles of welding are the same in repairs as in the construction. Additional complications arise since the welding has frequently to be carried out on a rigid structure.

Each repair must be considered separately. Damaged material to be cut away, avoid notch effect, corners should be cut round not square. Great attention must be given to the welding sequence allowing for contraction taking place on cooling of the weld metal. Successful repair work requires great attention to cutting, fitting and welding.

ROLLED AND BUILT SECTIONS.

Various rolled and built sections are used in ship construction and are illustrated. The type of section used depends on the degree of strength required. Built sections are used when a greater degree of strength is required than that obtained from rolled sections.

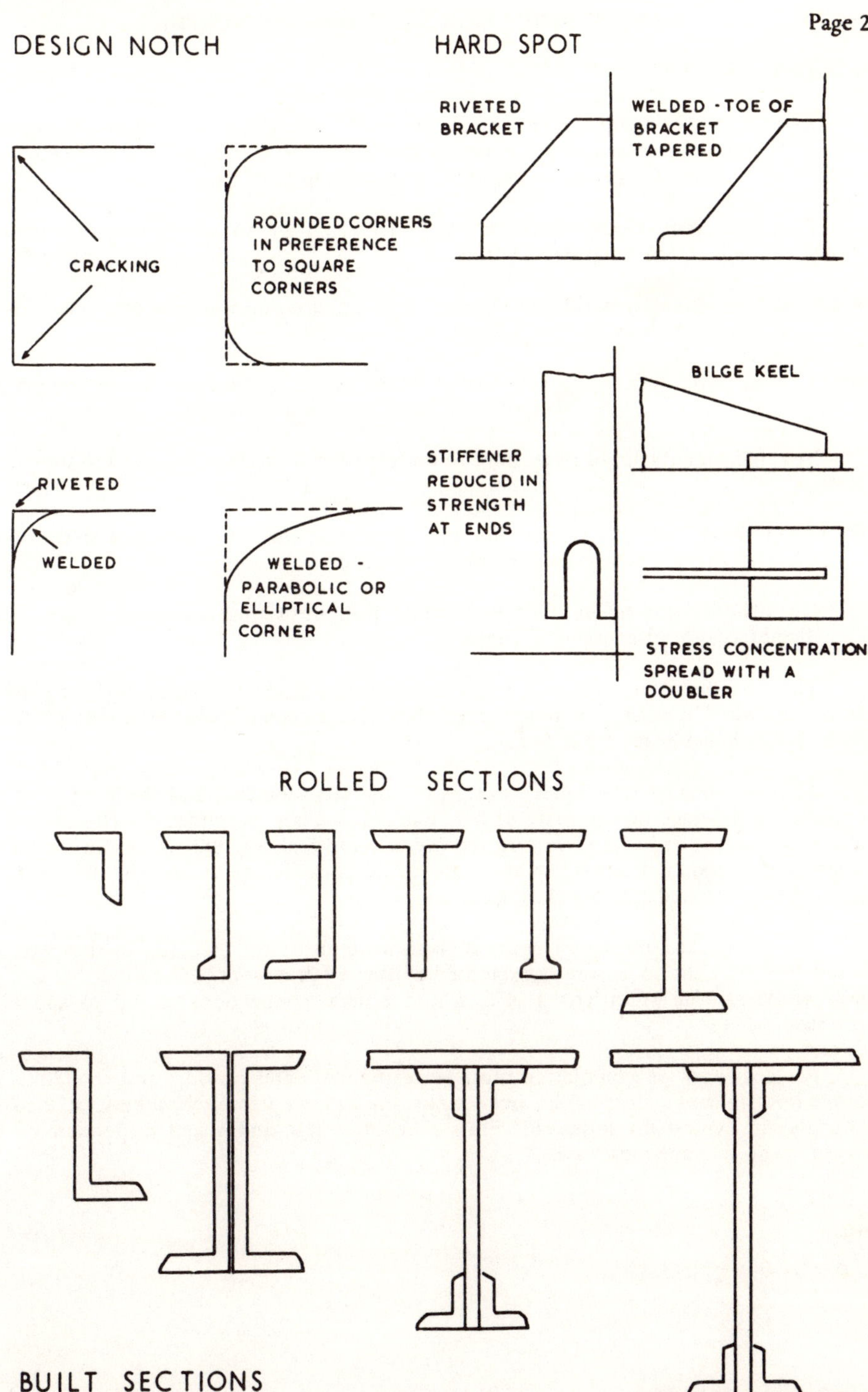
DESIGN NOTCH
HARD SPOT
RIVETED BRACKET
WELDED - TOE OF BRACKET TAPERED
CRACKING
ROUNDED CORNERS IN PREFERENCE TO SQUARE CORNERS
BILGE KEEL
STIFFENER REDUCED IN STRENGTH AT ENDS
RIVETED
WELDED
WELDED - PARABOLIC OR ELLIPTICAL CORNER
STRESS CONCENTRATION SPREAD WITH A DOUBLER
ROLLED SECTIONS
BUILT SECTIONS

RIVETING

Riveting has to a large extent been superseded by Welding and the latest Rules only refer to all welded construction. Parts of the vessel however may still be riveted, i.e. connection of deck stringer to sheerstrake, riveted straps.

The bars from which rivets are manufactured have to undergo certain tests as follows and illustrated on pages 15 and 17.

Tensile test – strength of 41–50 kg/mm^2 with an elongation of not less than 26% for steel bars.

Dump, compressive test – short lengths, when cold, to withstand being compressed to half their length without fracture.

Selected manufactured rivets have to undergo certain tests as follows and illustrated opposite :–

(1) Rivet shanks are to be bent cold and hammered until the two parts of the shank touch without fracture on the outside of bend.

(2) Rivet heads are to be flattened, while hot, until their diameter is 2½ times diameter without cracking at edges.

The size, diameter, of a rivet is governed by the thickness of the thickest plate or bar through which it passes, the length depending on the type of point required. Different types of points and heads are shown.

The punching of rivet holes weakens the metal so the centre of the holes must not be nearer to the edge of any plate or bar than 1½d. They are either punched from the 'faying' surface or drilled and any burred metal smoothed off. Where absolute water or oiltightness is required the rivet holes are to be countersunk as shown, the seams of plating being caulked on the same side.

The pitch, i.e. spacing of rivets is given in terms of the diameter of the rivet and should be sufficient to ensure satisfactory closing of the work. The pitch for oiltight work is 3½–4, for watertight 4–4½, whilst general connections are up to about 7 diameters.

Riveted work on completion must be tested. A water pressure test for tanks, i.e. double bottom tanks, deep tanks, peak tanks, etc., a high pressure hose test, i.e. shell and deck plating, watertight bulkheads etc., a hammer test and visual inspection of the riveted work in remaining cases.

RIVETING

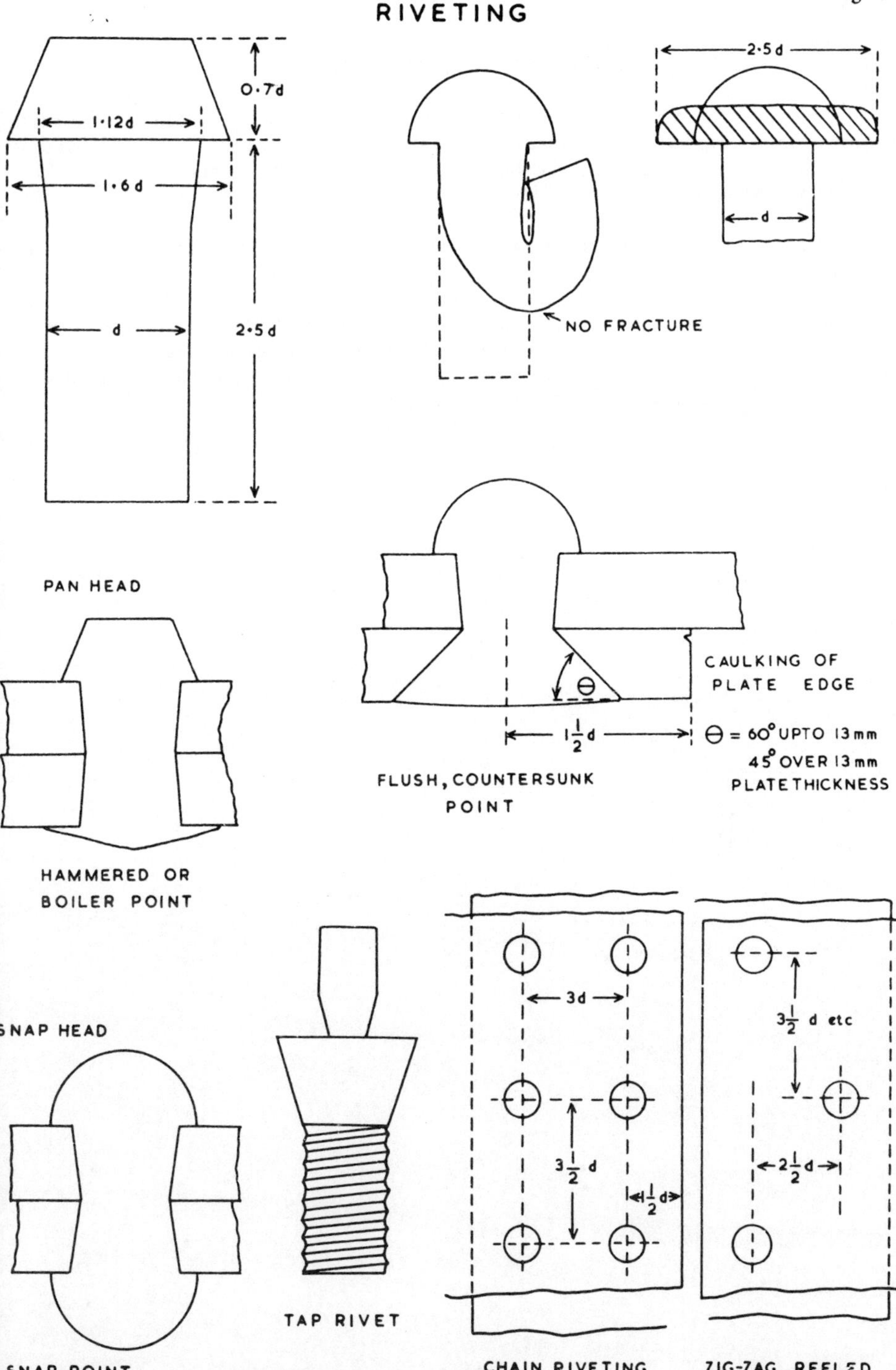
0·7d
1·12d
1·6d
d
2·5d
PAN HEAD
NO FRACTURE
2·5d
d
HAMMERED OR
BOILER POINT
CAULKING OF
PLATE EDGE
1½d
Θ = 60° UPTO 13 mm
45° OVER 13 mm
PLATE THICKNESS
FLUSH, COUNTERSUNK
POINT
SNAP HEAD
SNAP POINT
TAP RIVET
3d
3½d
1½d
CHAIN RIVETING
3½ d etc
2½d
ZIG-ZAG REELED
RIVETING

CURVES OF SHEARING FORCE AND BENDING MOMENT.

The shearing force and bending moments at sections along the length of a beam may be shown graphically by plotting the values of the shearing force and bending moment at various points along the beam. The resultant graphs may be straight lines or curves and show the variation of the stress along the beam.

Such curves are explained in the companion volume "Ship Stability Notes and Examples" by Kemp and Young.

The illustrations opposite show the associated curves of shear force and bending moments for a ship in the still water condition and when amongst waves, in the first instance the wave crests at each end and in the second instance the wave crests at each end and trough amidships, i.e. hogging and sagging.

TYPICAL STRENGTH CURVES

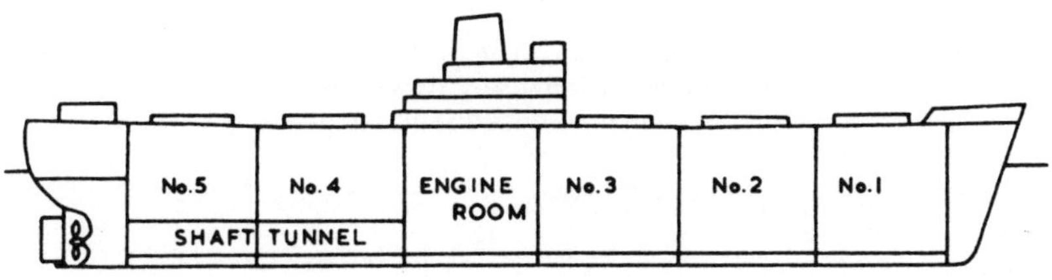

STILL WATER CONDITION

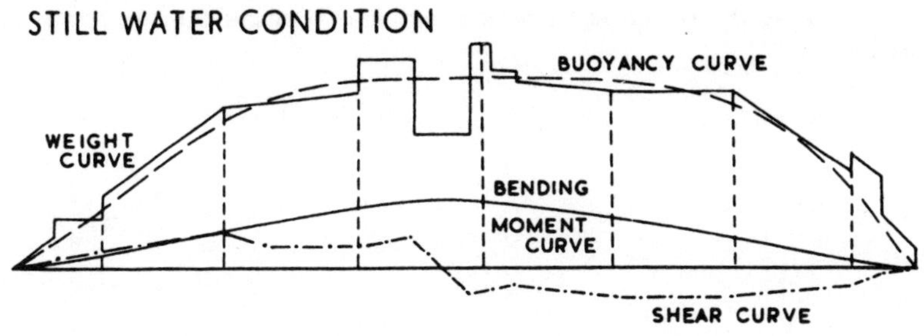

HOGGING CONDITION

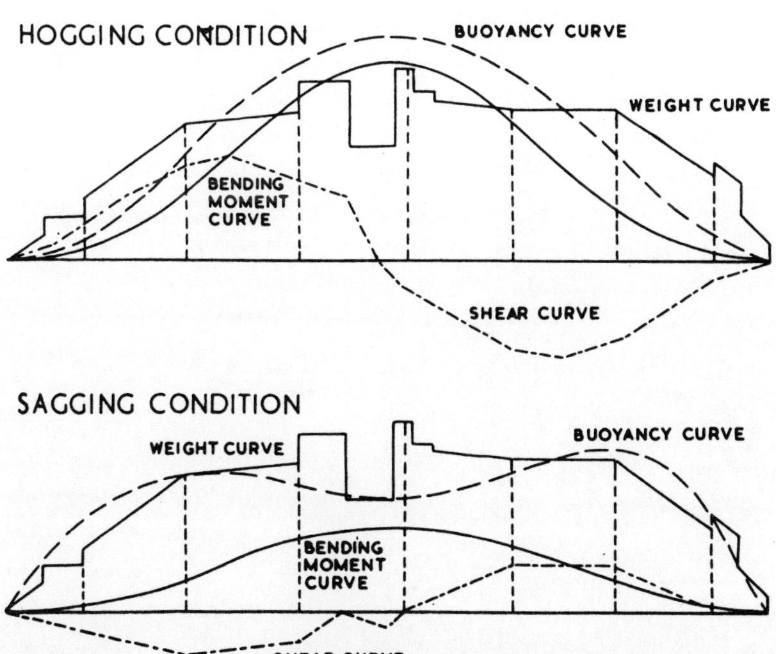

A ship at sea is subjected to a number of forces causing the structure to distort. Initially these may be divided into two categories, as follows :–

(1) Static forces – ship floating at rest in still water. Two forces acting, (a) weight of ship acting vertically downward and (b) water pressure acting perpendicular to outside surface of ship.

(2) Dynamic forces – ship in motion.

Six degrees of freedom, three linear and three rotational, are illustrated. When these motions are large then very large forces may be generated. Although the forces generated are usually of a local nature, i.e. heavy pitching resulting in pounding, they are liable to cause the structure to vibrate and thus transmit the stresses to other parts of the structure.

FORCES CAUSING STRESS

STATIC FORCES

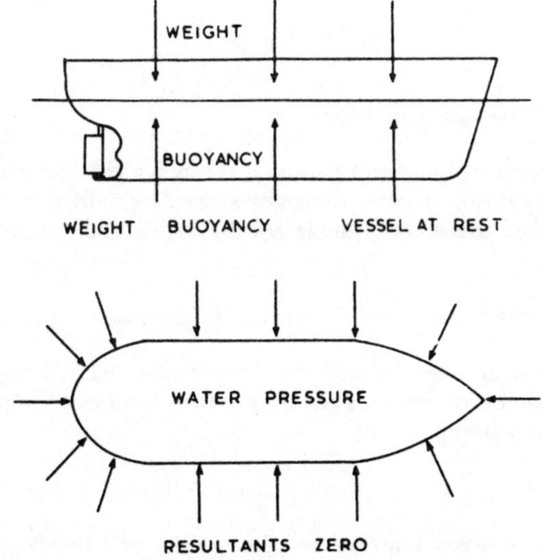

DYNAMIC FORCES

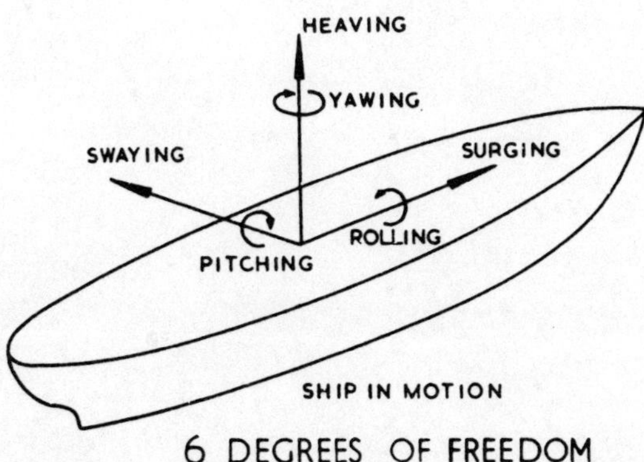

6 DEGREES OF FREEDOM

The separate stresses to which the ship's structure is subjected may be divided as follows :–

(1) Structural – those affecting the whole ship,

(2) Local – those affecting particular parts of the ship.

Structural Stresses.

(a) **Longitudinal Stresses in Still Water.**

Although the upthrust (buoyancy) is equal to the weight of the ship the distribution of weight and buoyancy is not uniform throughout the length of the vessel and differences (load) occur throughout the length, giving rise to tensile and compressive stresses away from the neutral axis.

(b) **Longitudinal Stresses in a Seaway causing Hogging and Sagging.**

When the ship is amongst waves the weight distribution remains unchanged but the distribution of buoyancy is altered. Similar conditions may be experienced or magnified by incorrect loading.

(c) **Shearing Stresses.**

The longitudinal stresses imposed by the weight and buoyancy distribution may give rise to longitudinal shearing stresses. The maximum shearing stress occurs at the neutral axis, a minimum at the deck and keel. Vertical shearing stresses may also occur.

STILL WATER CONDITION

LIGHTSHIP

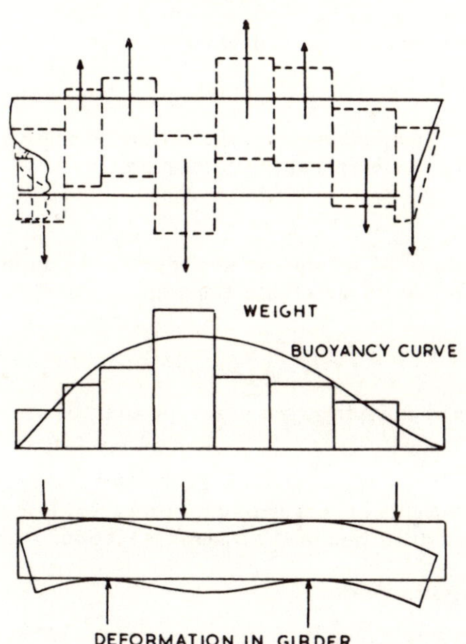

WEIGHT
BUOYANCY CURVE
DEFORMATION IN GIRDER

HOGGING

WEIGHT
BUOYANCY
BUOYANCY CURVE
WAVE CREST AMIDSHIPS
STILL WATER
TENSION
COMPRESSION

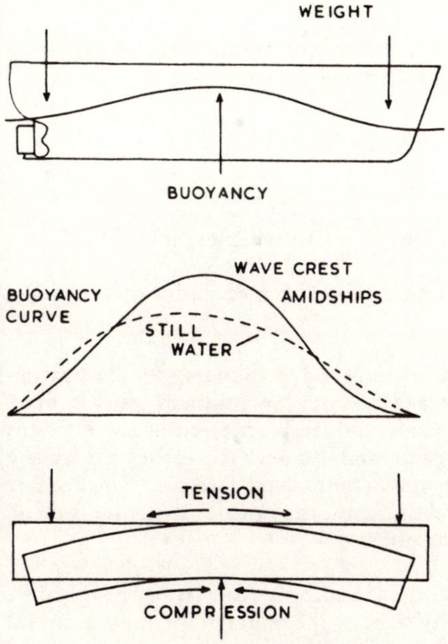

SAGGING

WEIGHT
BUOYANCY
BUOYANCY CURVE
STILL WATER
WAVE CREST AT ENDS
COMPRESSION
TENSION

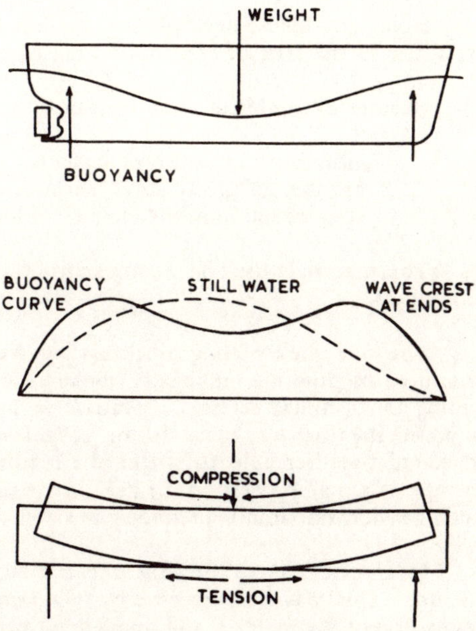

(d) **Racking.**

When a vessel is rolling the accelerations on the ship's structure are liable to cause distortions in the transverse section. The greatest effect is under light ship conditions.

(e) **Water Pressure.**

Water pressure acts perpendicular to the surface increasing with depth. The effect of this is to push the ship's sides in and the bottom up.

(f) **Drydocking.**

This has a tendency to set up the keel due to the upthrust of the keel blocks resulting in a change in the shape of the transverse section.

Localised Stresses.

(1) and (2) **Panting and Pounding,** see pages 44 and 46.

(3) **Localised Loading.**

Localised heavy weights i.e. engineroom, or localised loading of heavy cargo i.e. ore may give rise to localised distortion of the transverse section.

(4) **The Ends of Superstructures**

These may represent major discontinuities in the ship's structure giving rise to localised stresses which may result in cracking.

(5) **Deck Openings.**

Holes cut in the deck plating, i.e. hatchways, masts, etc. create areas of high local stress due to the lack of continuity created by the opening.

(6) **Further examples of local regions of stress are** :

vibration due to propellers
stresses set up by stays, shrouds etc.
stresses set up in the vicinity of hawspipes, windlass winches, etc.

The materials used in a ship's structure form a boxed shaped girder of very large dimensions.

The side shell plating, deck plating, hatch coamings, deck girders, double bottom structure, longitudinal bulkheads (tankers), deck and bottom longitudinals assist in overcoming longitudinal stresses. Transverse bulkheads and deep transverses are efficient in preserving the transverse form. Frames, beams and floors, etc. all being securely bracketed together help to stiffen the plating against compressive stresses. Since water pressure is a major stress on the hull, increasing with depth, the bottom structure is made much heavier, side framing reducing in size with height.

It is essential to prevent the various stresses causing deformation or possible fracture. This may be overcome by increasing the sizes of the material used, by a careful disposition of the material and by paying careful attention to the structural design.

RACKING

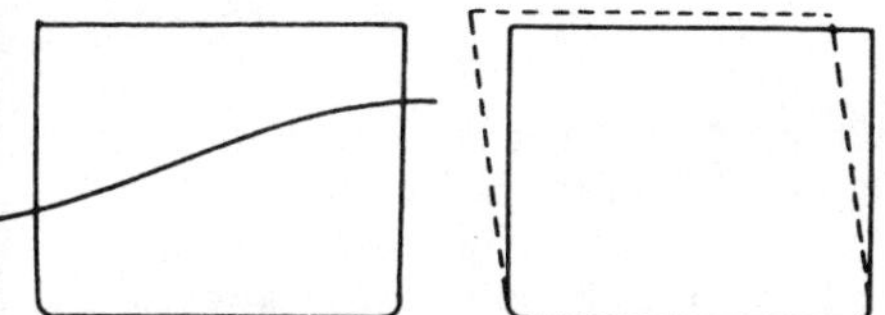

LOCALISED LOADING

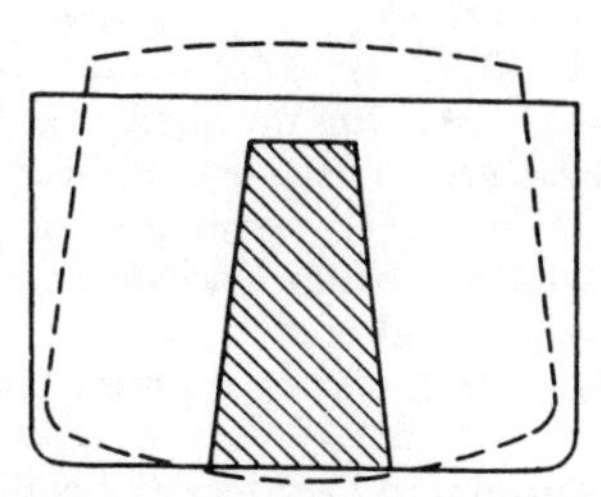

DISTORTION OF TRANSVERSE SECTION

DECK OPENINGS

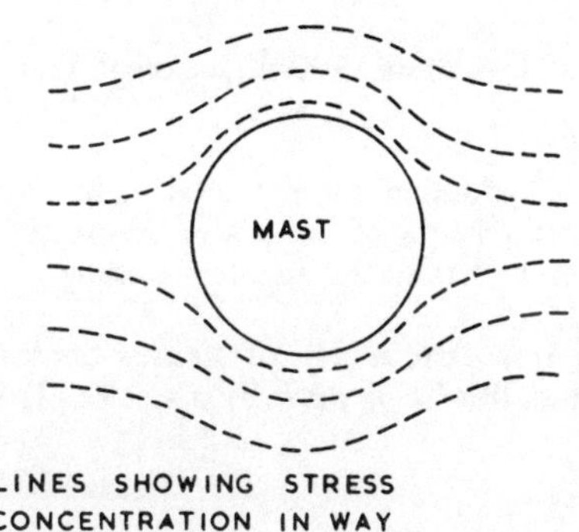

LINES SHOWING STRESS CONCENTRATION IN WAY OF OPENINGS

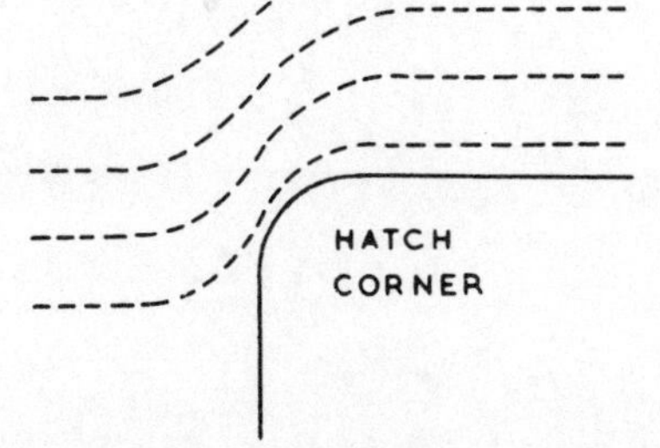

WATER PRESSURE

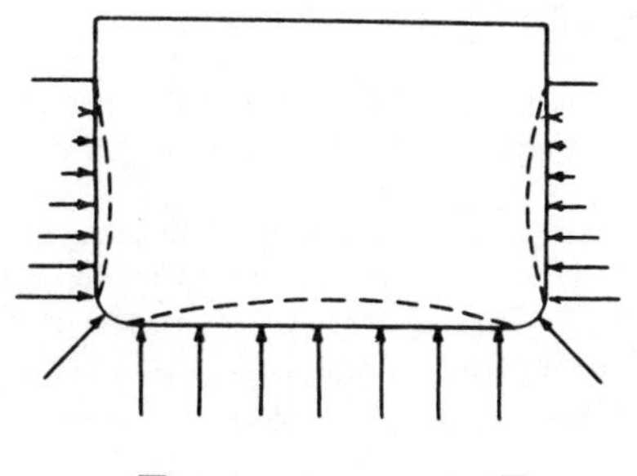

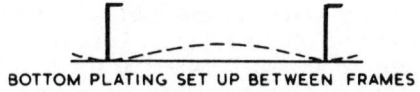

BOTTOM PLATING SET UP BETWEEN FRAMES

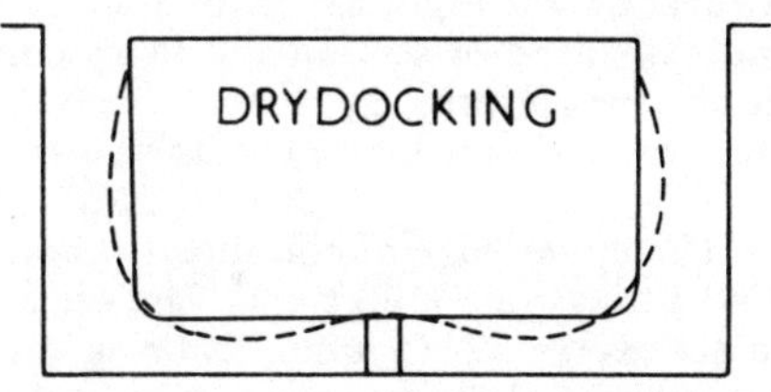

SUPERSTRUCTURE ENDS

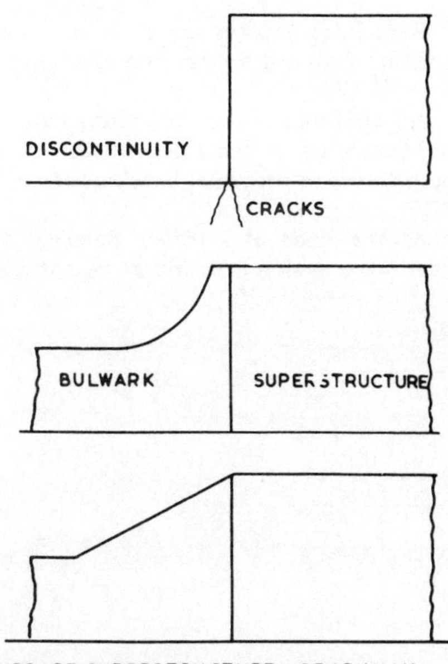

ENDS OF SUPERSTRUCTURE GRADUALLY TAPERED OFF NO ABRUPT CHANGES OF FORM

DECK AND SHELL PLATING

This is the most important part of a vessel's structure since it forms both a watertight skin and is the principle longitudinal strength member of the hull girder.

The plates are arranged in fore and aft lines around the hull, called Strakes, and for identification are lettered starting with the strake adjacent to the keel, this strake being 'A'. The separate plates in the strakes are numbered, usually from aft, thus plate 'C 12 port' will be the 12th plate from aft in the 3rd strake up from the keel on the port side.

The thickness of the plating depends, in general, on the length of the vessel and frame spacing. The midship thickness is to be maintained for 0.4 L amidships and tapered gradually to an end thickness at 0.1 L at the deck and 0.075 L for the shell from the ends. Special attention to thickness is required when decks are to carry excess loads, to structural details especially with respect to hatch corners and deck openings. Abrupt changes of shape or section and sharp corners are to be avoided. Where decks are coated with an anti-corrosive paint and a wood sheathing or composition is effectively secured to the deck then a reduction of up to 10% in the thickness of deck plating is permitted.

The upper edge of the sheerstrake is to be dressed smooth and kept free of isolated welded fittings or connections. The width amidships is not to be less than 150D mm. but need not exceed 2135 mm., D being the moulded depth in metres. Where the sheerstrake is rounded the radius is not to be less than 15 times the thickness of the plate.

The width of the keel is not to be less than 70 B mm but need not exceed 1800mm, B being the moulded breadth. Its thickness is to be not less than 6 + 0.1 L mm.

Remaining strakes are to have a breadth not less than 1500 mm except that the bilge strake shall not be less than 1800 mm.

All openings in the shell plating are to have well rounded corners. Plating in way of hawsepipes is to be increased in thickness and the thickness of the plates connected to the sternframe or propeller brackets are to be 50% greater than the adjacent plating.

At the ends of a vessel, particularly at the bow, the width of strakes decreases and it is often desirable to merge two strakes into one, this being done by a stealer plate.

SHELL EXPANSION PLAN

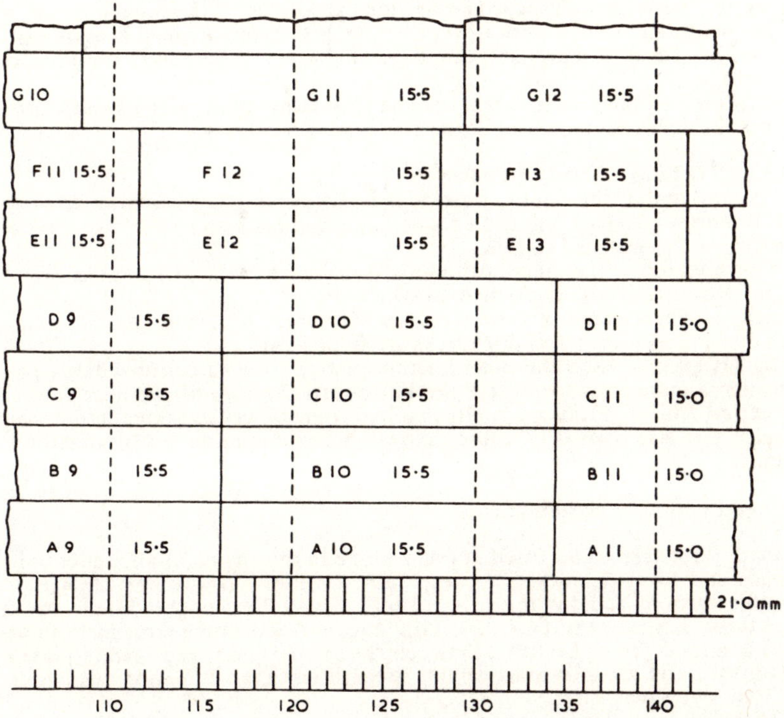

G 10
G 11 15·5
G 12 15·5
F 11 15·5
F 12 15·5
F 13 15·5
E 11 15·5
E 12 15·5
E 13 15·5
D 9 15·5
D 10 15·5
D 11 15·0
C 9 15·5
C 10 15·5
C 11 15·0
B 9 15·5
B 10 15·5
B 11 15·0
A 9 15·5
A 10 15·5
A 11 15·0
21·0mm
110 115 120 125 130 135 140

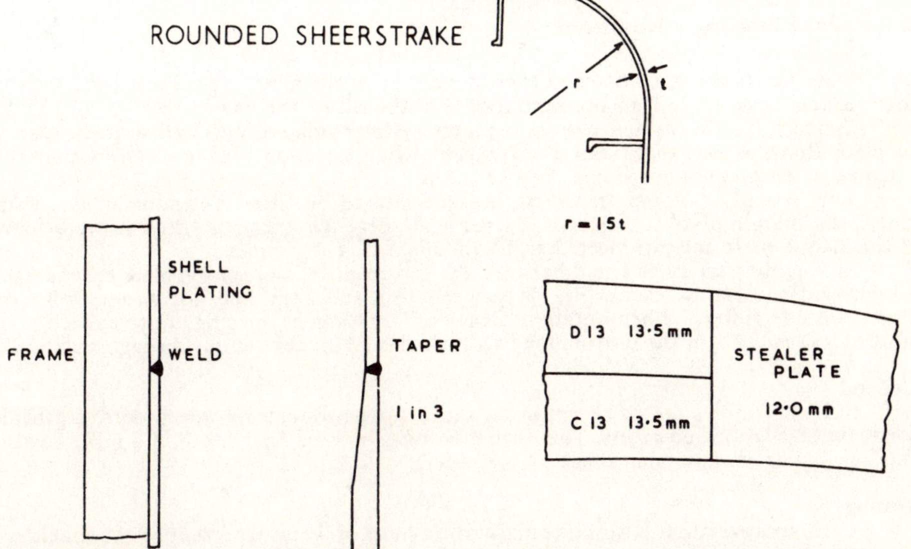

THE KEEL, CENTRE GIRDER AND CENTRE LINE STRAKE of tank top plating form a very strong I shaped girder, the 'backbone' of the vessel.

The width and thickness of the keel strake is to be maintained over its whole length. See page 36. In an average size cargo vessel these will be about 1420mm and 20mm respectively.

The keel is joined to the stem bar and sternframe by the SHOE and COFFIN plate which are the first and last plates of the flat plate keel.

DOUBLE BOTTOM CONSTRUCTION

Framing within the double bottom is to be either longitudinal or traverse. Vessels over 120 metres in length are to be framed longitudinally and when the notation "Heavy or Ore Cargoes" is assigned to a vessel.

The thickness of the inner bottom plating is to be calculated and increased when the notation "Heavy Cargoes" is assigned, or where there is no ceiling fitted in the square of the hatch, or where cargo is to be regularly discharged by grabs. A minimum thickness is given in the Rules when fork lift trucks are to be used.

In Passenger Vessels the inner bottom plating is to be continued out to the ship's side in such a manner as to protect the bottom to the turn of the bilge (Merchant Ship Construction Rules). Drainage is effected by means of wells situated in the wings, having a capacity not less than 0.17 cubic metres and extending to not nearer the shell tthan 460mm.

Transverse Framing – Requirements

Plate floors are to be fitted at every frame in the engine room, under boilers, under bulkheads and toes of brackets to deep tank stiffeners, in way of a change of depth in the double bottom and for the forward 0.25 L (see Pounding). Elsewhere the spacing of plate floors is not to exceed 3.05m with bracket floors at the remaining frames.

Side girders are to be fitted between the centre girder and margin plate extending as far forward and aft as is practicable. Breadth less than 20m, one side girder; breadth above 20m, two side girders.

The unsupported span of the frames in bracket floors is not to exceed 2.5m. Breadth of brackets attaching the frames to the centre girder and margin plate is to be ¾ depth of centre girder. The brackets are to be flanged on their unsupported edge.

Longitudinal Framing – Requirements

Plate floors are to be fitted at every frame under the main engines and the foremost shaft tunnel bearing and at alternate frames outboard of the engine seating, also under boiler bearers, bulkheads and toes of brackets to deep tank stiffeners. Elsewhere spacing of plate floors is not to exceed 3.7m except when notation 'Heavy or Ore Cargoes' is assigned when maximum spacing is to be 2.5m.

Between plate floors transverse brackets are to be fitted extending from centre girder and margin plate to adjacent longitudinal. Brackets are to be fitted at every frame at the margin plate and not more than 1.25m apart at centre girder.

Side girders are to be fitted between the centre girder and margin plate extending as far forward and aft as practicable. B between 14m and 21m, one side girder; B over 21m, two side girders. When notation 'Heavy or Ore Cargoes' spacing not to exceed 3.7m. Where L exceeds 215m the bottom longitudinals should be continuous through bulkheads.

General

Sufficient holes are to be cut in the inner bottom floors and side girders to provide adequate ventilation and access. Their size should not exceed 50% depth of double bottom and be circular or eliptical in shape.

Testing

Each compartment is to be tested with a head of water representing the maximum pressure which could be experienced in service or alternatively air pressure testing may be used.

DOUBLE BOTTOM CONSTRUCTION

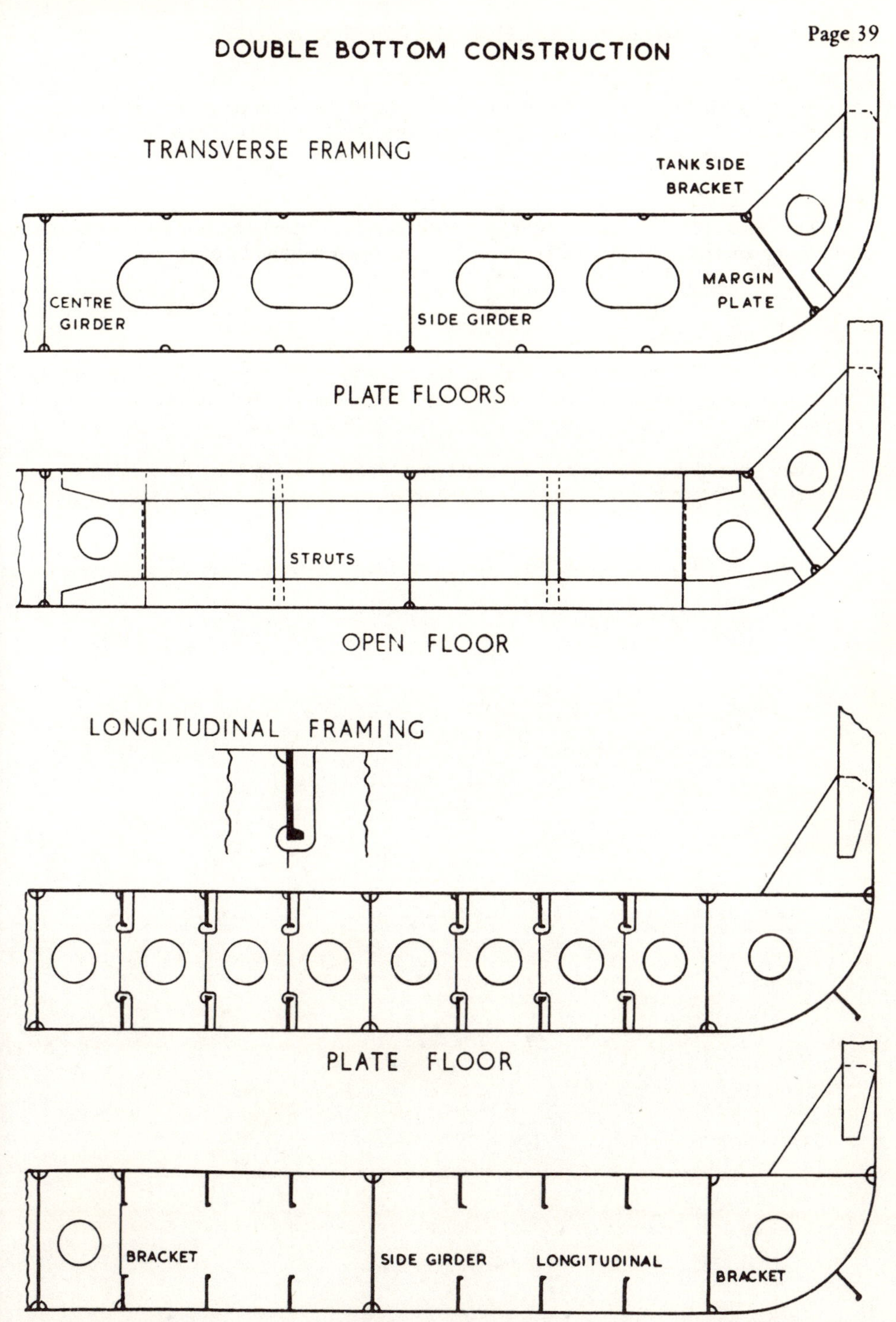

A DUCT KEEL or PIPE TUNNEL, consisting of 'twin' centre girders, is frequently fitted forward of the engine room, pipe lines being led through it. These 'ducts' may be fitted between the centre line and ship's side and are then referred to as pipe tunnels.

The sides are not to be more than 1.83m apart and the inner bottom and bottom shell plating are to be suitably stiffened so as to maintain continuity of strength.

BILGE KEELS

Bilge keels are fitted at the turn of the bilge to resist rolling; they also improve slightly the steering qualities of the vessel. They usually extend over the midship one-third or half length.

They should be attached to a continuous flat bar which may be welded to the shell plating. The ends are to be gradually tapered and should not end on an unstiffened panel.

Today, in many vessels, stabilizers are being fitted in lieu of bilge keels.

DUCT KEEL

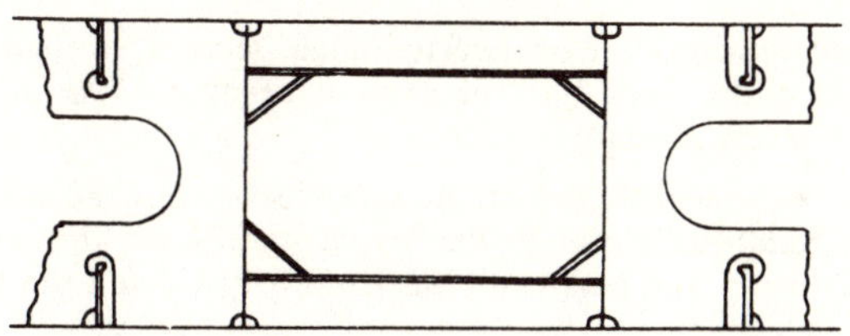

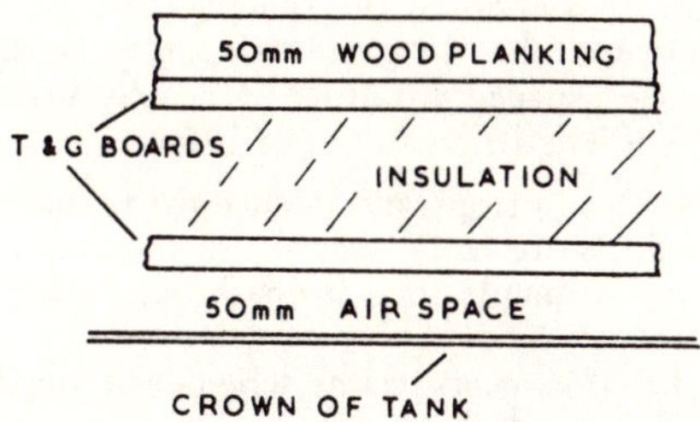

BILGE KEEL

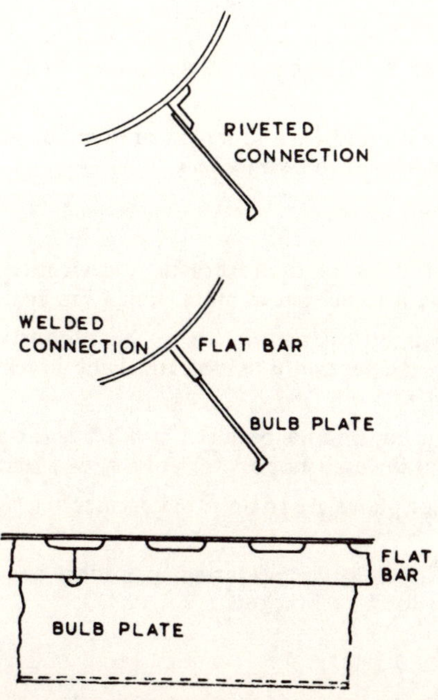

FRAMES, BEAMS AND LONGITUDINALS.

These are usually of bulb bar or inverted angle section. Longitudinal or transverse framing may be used at the deck but for ships exceeding 120m in length longitudinal framing is to be used at the strength deck.

The spacing of transverse frames at the sides forward of the collision bulkhead and abaft the after peak bulkhead is not to exceed 610mm, from the collision bulkhead to 0.2L from forward 700mm, and between 0.2L from forward and the after peak bulkhead 1000mm.

Transverse frames are not usually continuous, scantlings increasing with depth. For identification purposes frames are usually numbered from aft forward commencing at the transom floor. Frames abaft the transom floor are usually lettered.

Where longitudinal framing is adopted the spacing depends on the section modulus of the bulb bar, inverted angle etc. The longitudinals are to be supported at the side by webs, at the decks by transverses, spaced not more than 3.7m apart when the length does not exceed 185m and L/50m for lengths greater than 185m.

Where L exceeds 215m the bottom and deck longitudinals should be continuous through bulkheads. Longitudinals are to be secured to the bulkheads in such a manner as to maintain direct continuity of longitudinal strength, see Tankers, page 110.

Deck beams are to be fitted at every frame when deck longitudinals are not fitted. Their function is to support the decks carrying cargo or other loads and to act as struts holding the sides of the vessel apart against the inward pressure of the sea.

BEAM KNEES are fitted to provide an efficient connection between the frames and beams. They have a small amount of resistance against racking stresses. The rules govern the length of overlap, welding area, width of flange (minimum 50mm), size and thickness.

Tank Side Bracket

The lower end of the frame is to be connected to the tank top or margin plate by a bracket as illustrated.

The same rules, with reference to length of overlap, welding area, size of flange, etc., apply to tank side brackets as to beam knees.

DECK GIRDERS

In the forward 0.075L of the forecastle and weather decks the longitudinal girders supporting beams are not to be spaced more than 3.7m apart.

Girders of unsymmetrical section are to be supported by tripping brackets at alternate beams. When the section is symmetrical the brackets may be four frame spaces apart.

If a girder is not continuous between bulkheads the part which is under the deck is to be extended beyond the end support for at least two frame spaces before tapering off.

Horizontal gusset plates are to be fitted connecting side girders to hatch end beams, see illustration.

Weather deck hatch coamings acting as girders should be extended beyond the hatchway ends in the form of brackets.

BEAM KNEES

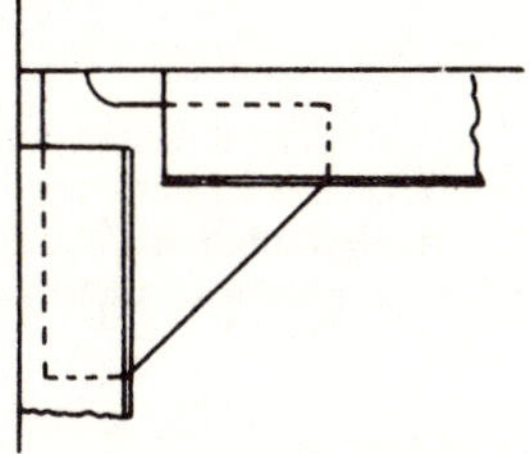

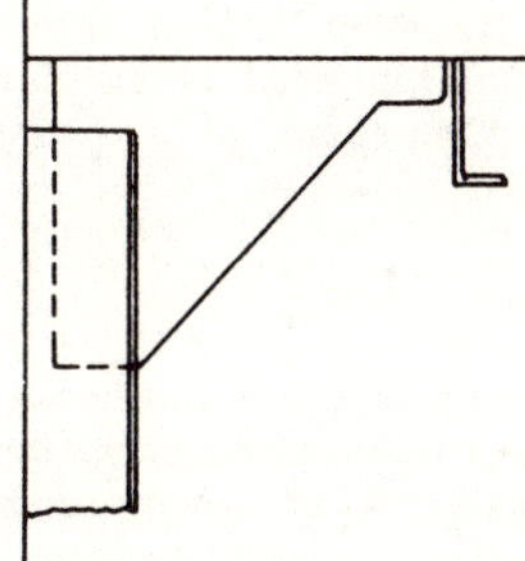

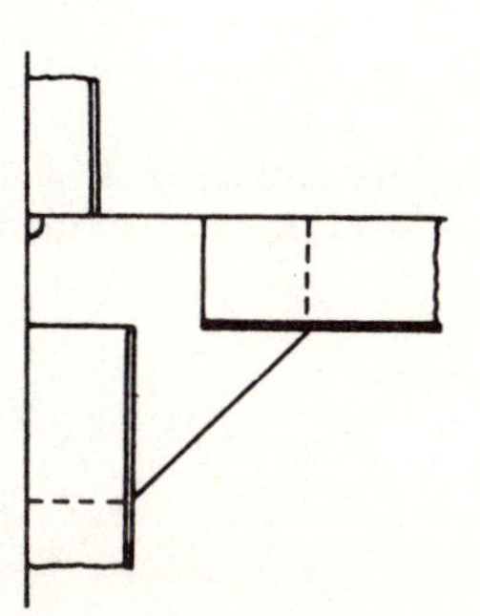

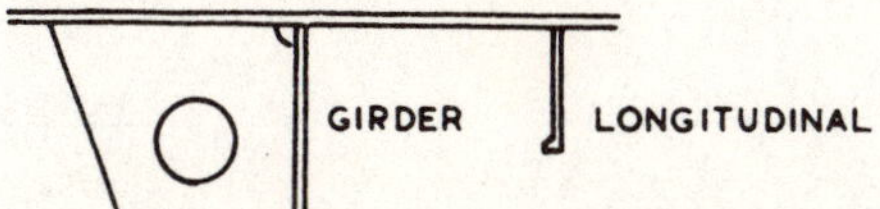

TRIPPING BRACKET AT ALTERNATE BEAMS OR EQUIVALENT (LONG. FRAMING)

TANKSIDE BRACKETS

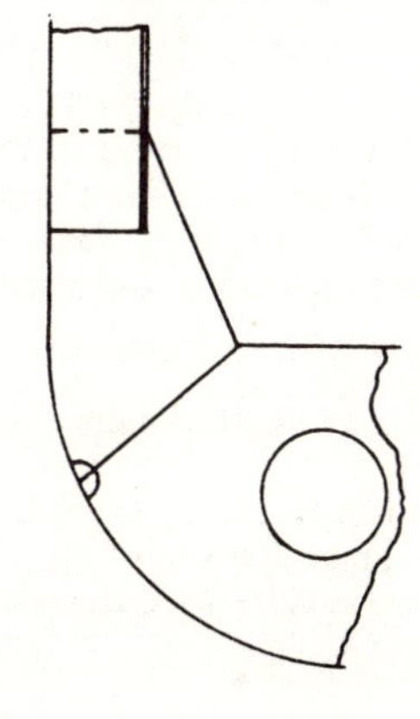

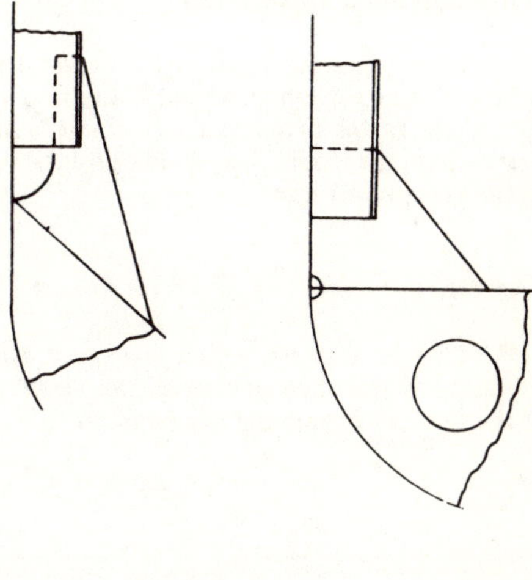

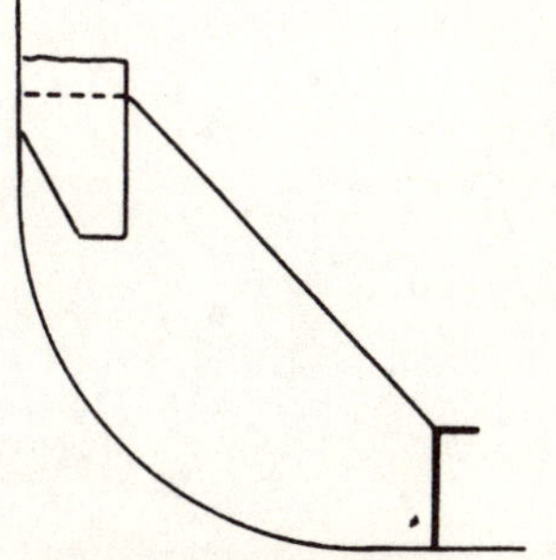

TANKSIDE BRACKET TANKER

POUNDING

Heavy pitching assisted by heaving as the whole vessel is lifted in a seaway may subject the forepart to severe blows from the sea. The greatest effect is experienced in the light ship condition. To compensate for this the bottom is strengthened from 0.05L to between 0.25L and 0.3L from forward depending on the block coefficient, unless the ballast draught forward is over 0.04L.

Bottom Framed Longitudinally

Longitudinals to be spaced 1000mm apart between 0.2L and 0.3L from forward and 700mm apart between 0.2L from forward and the collision bulkhead. Plate floors are to be fitted on alternate frames, side girders not more than 2.1m apart.

Bottom Framed Transversely

Frame spacing abaft 0.2L from forward is not to exceed 1000mm and between 0.2L and the collision bulkhead 700mm. Forward of the collision bulkhead 610mm. Plate floors are to be fitted at every frame. Intercostal side girders are to be not more than 2.2m apart with half height side girders not more than 1.1m apart, the girders extending as far as is practicable.

PANTING

This is a stress which occurs at the ends of a vessel due to variations in water pressure on the shell plating as the vessel pitches in a seaway. The effect is accentuated at the bow when making headway.

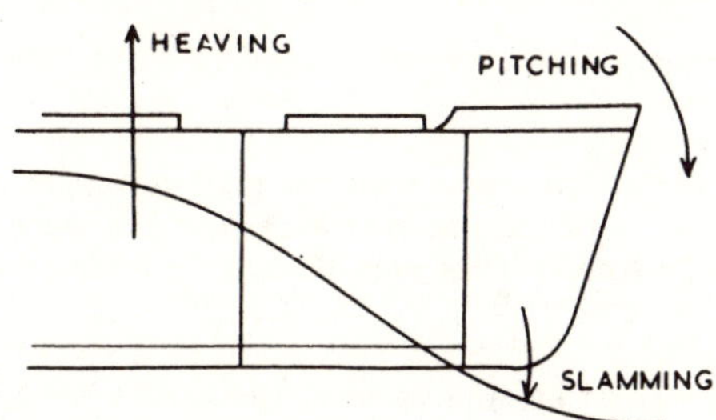
POUNDING
HEAVING
PITCHING
SLAMMING

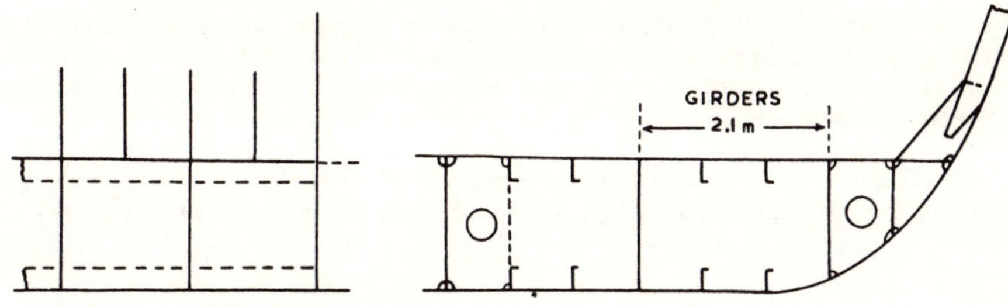
STRENGTHENING OF BOTTOM FORWARD
GIRDERS
2.1 m
PLATE FLOORS AT ALTERNATE FRAMES

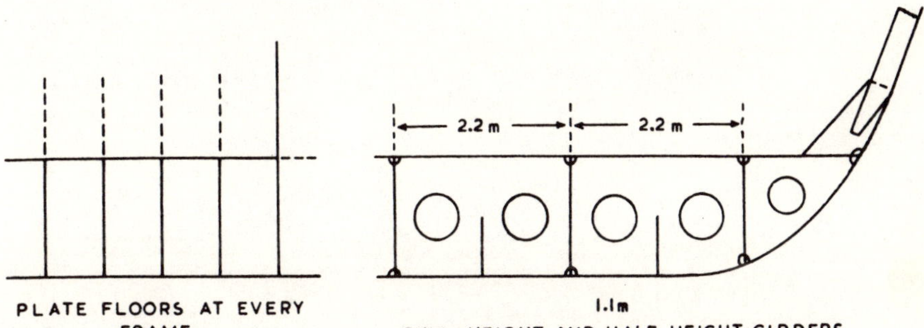
2.2 m
2.2 m
PLATE FLOORS AT EVERY FRAME
1.1m
FULL HEIGHT AND HALF HEIGHT GIRDERS

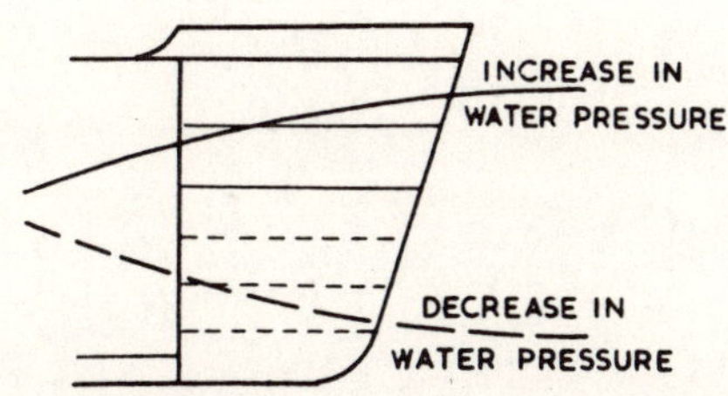
PANTING
INCREASE IN WATER PRESSURE
DECREASE IN WATER PRESSURE

Panting arrangements are to extend to 0.15L from forward and abaft the after peak bulkhead.

Tiers of beams spaced not more than 2m apart vertically are to be fitted at alternate frames in the forepeak or below the lowest deck above the waterline if the forepeak is small. Alternatively perforated flats may be fitted in lieu of panting beams 2.5m apart vertically.

Tiers of beams are to be supported at the centre line by a partial wash bulkhead or pillars. Beams are to be bracketed to frames and the frames to which no beams are attached are to be bracketed to the stringer. Stringer plates attached to the shell are to be fitted at each tier of beams.

Abaft the collision bulkhead intercostal side stringers having the same depth as the frames are to be fitted in line with those forward of the collision bulkhead and are to extend aft for 0.15L from the fore end. Stringers may be omitted if the shell plating is of increased thickness.

Abaft the after peak bulkhead the structure is to be efficiently stiffened by deep floors and tiers of beams in association with stringers spaced 2.5m apart vertically.

ARRANGEMENT FORWARD OF COLLISION BULKHEAD

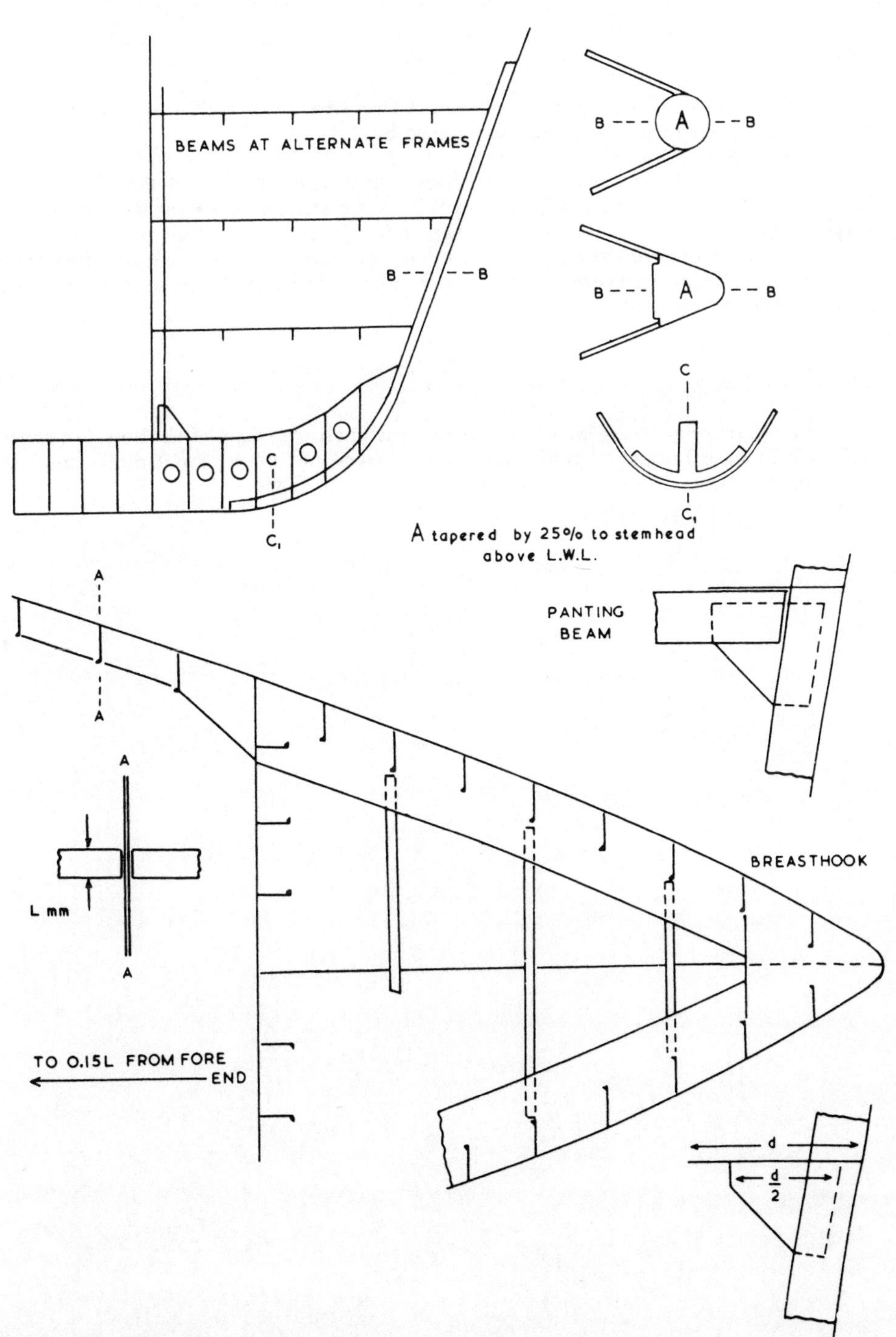

STEMS

Above the load waterline the cross sectional area of bar stems may be tapered to the stem head where the area may be reduced by 25%.

Where the stem is constructed of shaped plates the thickness of the plating above the load waterline may be gradually tapered to the stem head where it may have the same thickness as the shell plating at the ends. Plate stems are to be supported by horizontal webs between the decks and below the lowest deck the unsupported length of stem plates is not to exceed 1.5m. Where the curvature of the plate is large a centre line web may be required.

BULBOUS BOWS

Constructional arrangements are dependent upon the shape of the bulb. In general they are to have horizontal diaphragm plates spaced not more than 1 metre apart and when bulb is large a centre line wash bulkhead is to be fitted.

PANTING - ALTERNATIVE ARRANGEMENT

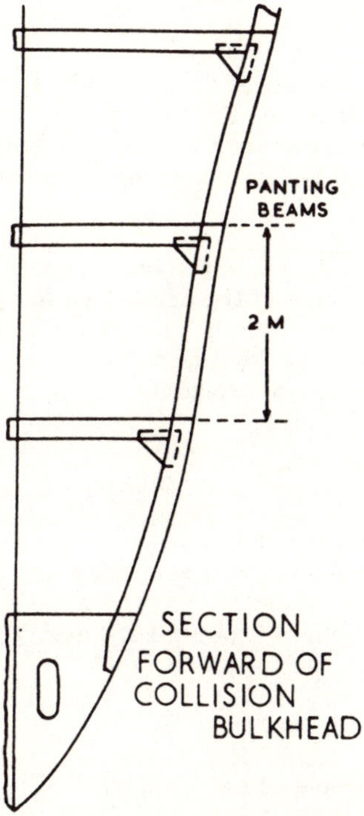

PANTING BEAMS

2 M

SECTION FORWARD OF COLLISION BULKHEAD

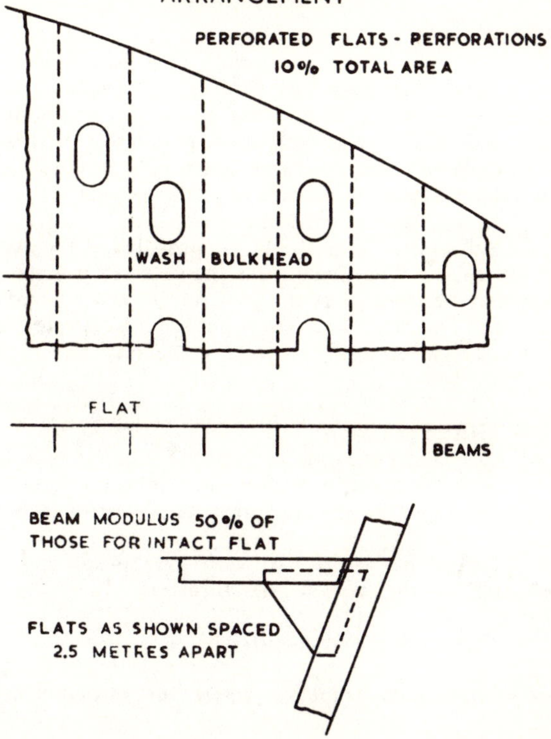

PERFORATED FLATS - PERFORATIONS 10% TOTAL AREA

WASH BULKHEAD

FLAT

BEAMS

BEAM MODULUS 50% OF THOSE FOR INTACT FLAT

FLATS AS SHOWN SPACED 2.5 METRES APART

BULBOUS BOW

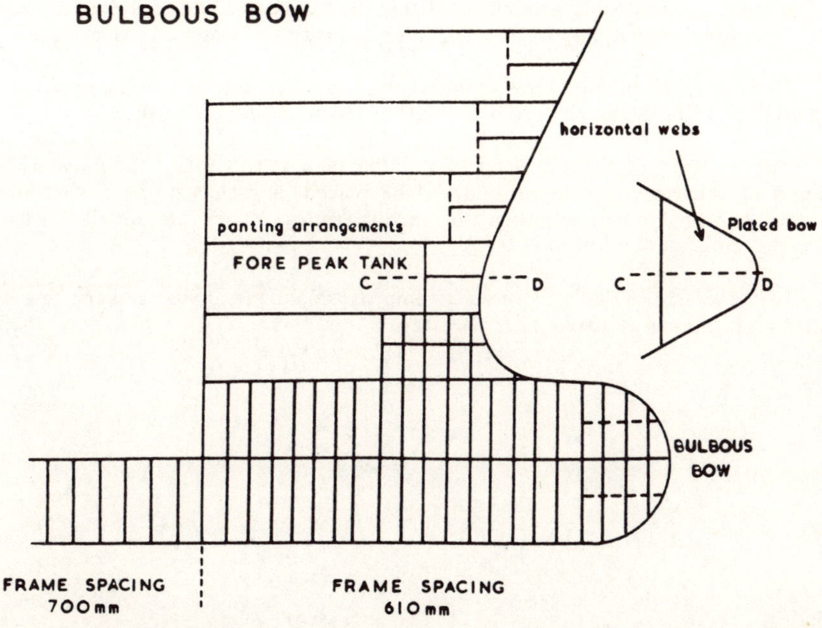

horizontal webs

Plated bow

panting arrangements

FORE PEAK TANK

C D

C D

BULBOUS BOW

FRAME SPACING 700mm

FRAME SPACING 610mm

HAWSE PIPES

Hawse pipes are cast to a diameter which should not be less than nine times the diameter of the cable they are to carry. A doubler is fitted to the shell plating and deck plating in way of each hawse pipe or the plating is increased in thickness. If a frame or beam is cut additional strengthening is required. Generally the axis of the hawsepipe should not exceed 45 degrees from the vertical.

Substantial flanges must be provided at the deck and shell to withstand the chafing of the cable. The illustration shows a heavy metal slab fitted at the deck. The hawse pipe may be inserted at the shell and drawn upwards into position or inserted from the deck and lowered. The latter method is illustrated and has the advantage that the shell flange, which receives most wear, is the loose one and can easily be replaced.

BOW THRUSTERS

With the great increase in the size of the very large tankers, bulk carriers, container-ships, passenger vessels, etc., directional control at low speed is of primary importance.

Directional control at very low speeds and especially when berthing has been obtained by the use of bow thrusters.

These units may consist of :–

(a) a shrouded propeller, where the shroud is movable and acts as a rudder,

(b) a transverse tunnel or duct through the ship near the bow in the narrow forward section. A reversible propeller is fitted on the centre line of the tunnel which acts as a pump discharging or throwing large quantities of water to either side.

Modern developments have replaced the propeller, in many cases, by an ejector, or separate high velocity nozzles are fitted within the main line of the shell.

The position of bow thrusters should be such as to control the movement of the forepart of the vessel. They should not be placed in such a position that the orifice becomes damaged when working the anchor cables. They are usually situated just abaft the collision bulkhead.

In some large vessels in order to improve manoeuvrability at very low speeds a similar arrangement is fitted near the stern.

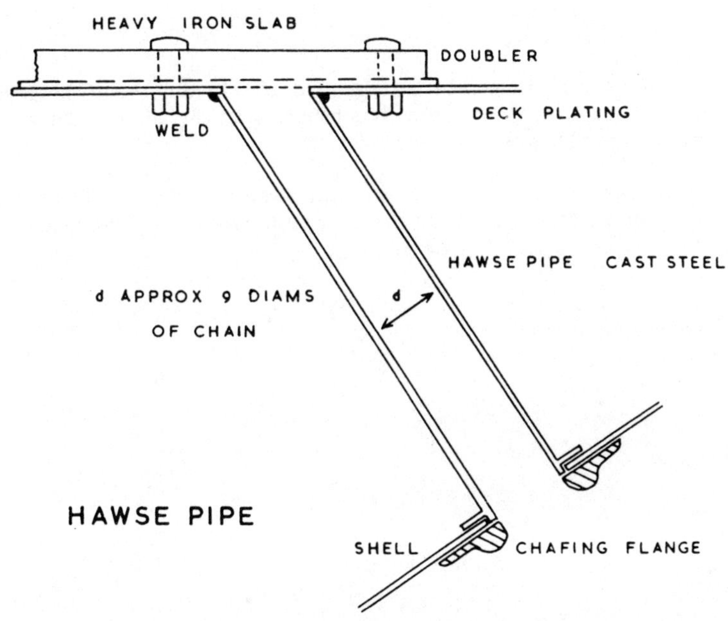
HEAVY IRON SLAB
DOUBLER
DECK PLATING
WELD
HAWSE PIPE CAST STEEL
d APPROX 9 DIAMS OF CHAIN
d
HAWSE PIPE
SHELL
CHAFING FLANGE

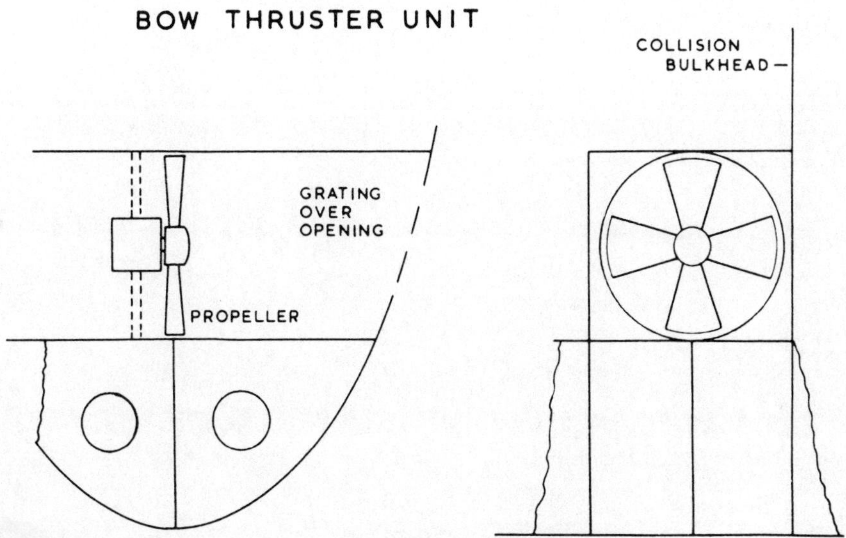
BOW THRUSTER UNIT
COLLISION BULKHEAD
GRATING OVER OPENING
PROPELLER

The illustration shows the general arrangement aft. In this figure the engineroom is situated aft. The transom floor is usually stronger than the adjacent floors and it is to this floor that the rudder post is connected.

Abaft the transom floor centre line and side girders are fitted. Transverse floors and frames are spaced 610mm apart abaft the after peak bulkhead. Panting arrangements within this section are as detailed on page 46.

GENERAL ARRANGEMENT AFT

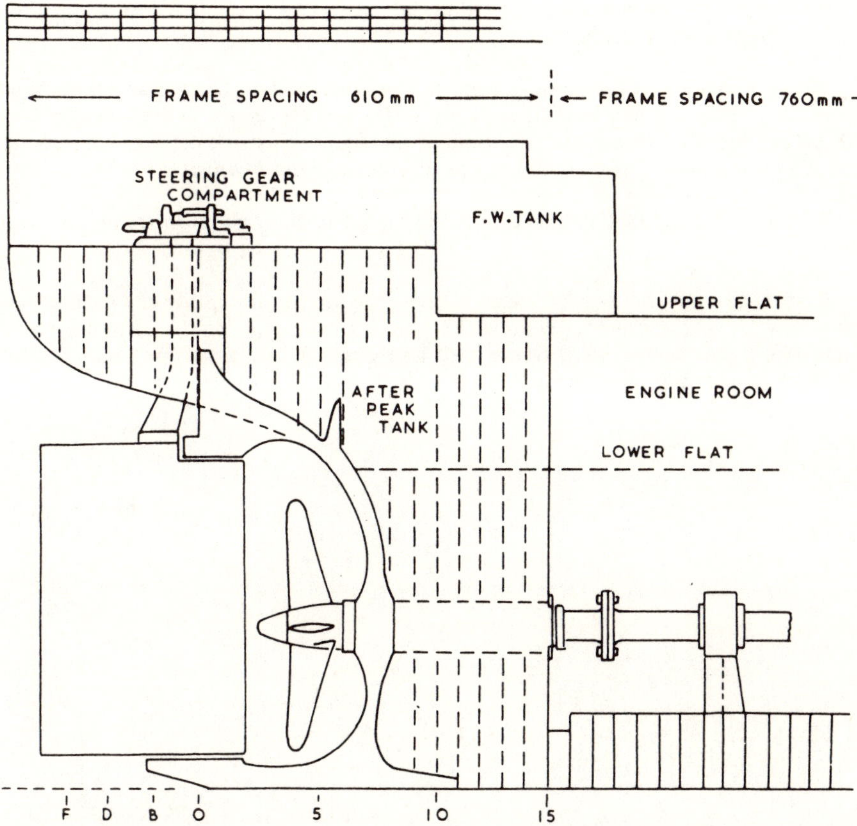

FRAME SPACING 610mm
FRAME SPACING 760mm
STEERING GEAR COMPARTMENT
F.W.TANK
UPPER FLAT
ENGINE ROOM
AFTER PEAK TANK
LOWER FLAT
F
D
B
O
5
10
15

RIVETED SCARPH

13%t
3d
d
t

AT-LEAST 4 ROWS OF RIVETS

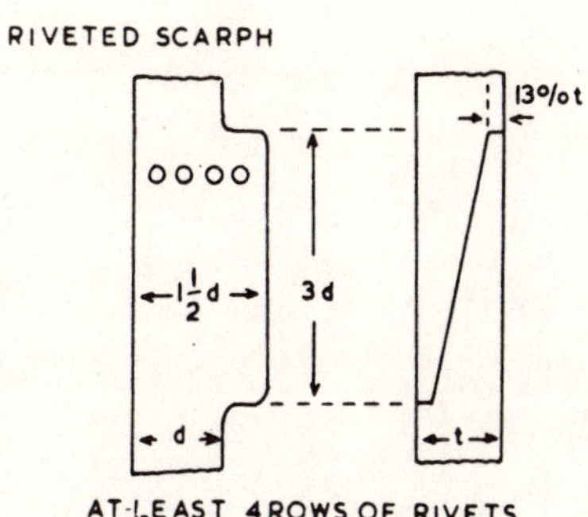

STERN FRAMES

Stern frames may be cast, forged or fabricated from steel plate. In the case of cast or forged steel stern frames they may be in one piece or in two or more sections riveted or thermit welded together.

Where a riveted connection is used the two sections of bar are scarphed together and the Rules give the length of the scarph as 3D and the depth as 1½D, where D is the depth of the bar used in the construction of the frame, this is illustrated on page 53. A scarph fitted in a rudder post should not be above the highest gudgeon.

The use of a welded connection is illustrated in the cast stern frame and semi or balanced rudder.

Cast steel and fabricated stern frames are to be strengthened at intervals by transverse webs. All stern frames are to be efficiently attached to the adjoining structure and the lower part of the stern frame is to be extended forward to provide an efficient connection to the flat plate keel.

CAST STEEL STERN FRAMES

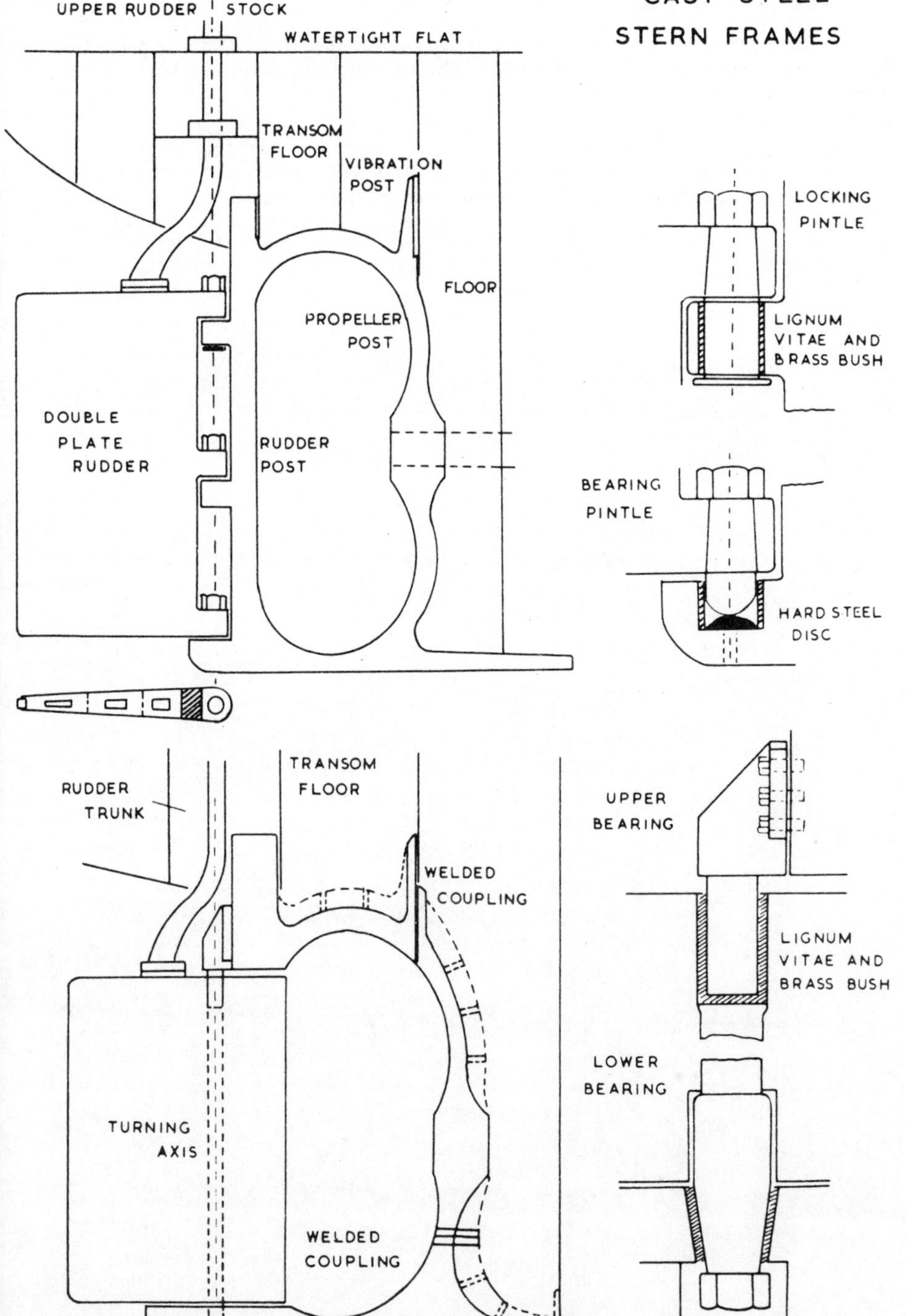
UPPER RUDDER STOCK
WATERTIGHT FLAT
TRANSOM FLOOR
VIBRATION POST
FLOOR
PROPELLER POST
DOUBLE PLATE RUDDER
RUDDER POST
LOCKING PINTLE
LIGNUM VITAE AND BRASS BUSH
BEARING PINTLE
HARD STEEL DISC
RUDDER TRUNK
TRANSOM FLOOR
WELDED COUPLING
UPPER BEARING
LIGNUM VITAE AND BRASS BUSH
LOWER BEARING
TURNING AXIS
WELDED COUPLING

With larger stern frames there is a tendency for the whole stern or propeller post and adjacent sections to be fabricated. This is illustrated with accompanying cross sections in way of various frames.

Notice that where a balanced rudder is fitted the rudder post is omitted and the unsupported sole piece is then of a more substantial construction.

FABRICATED STERN FRAME

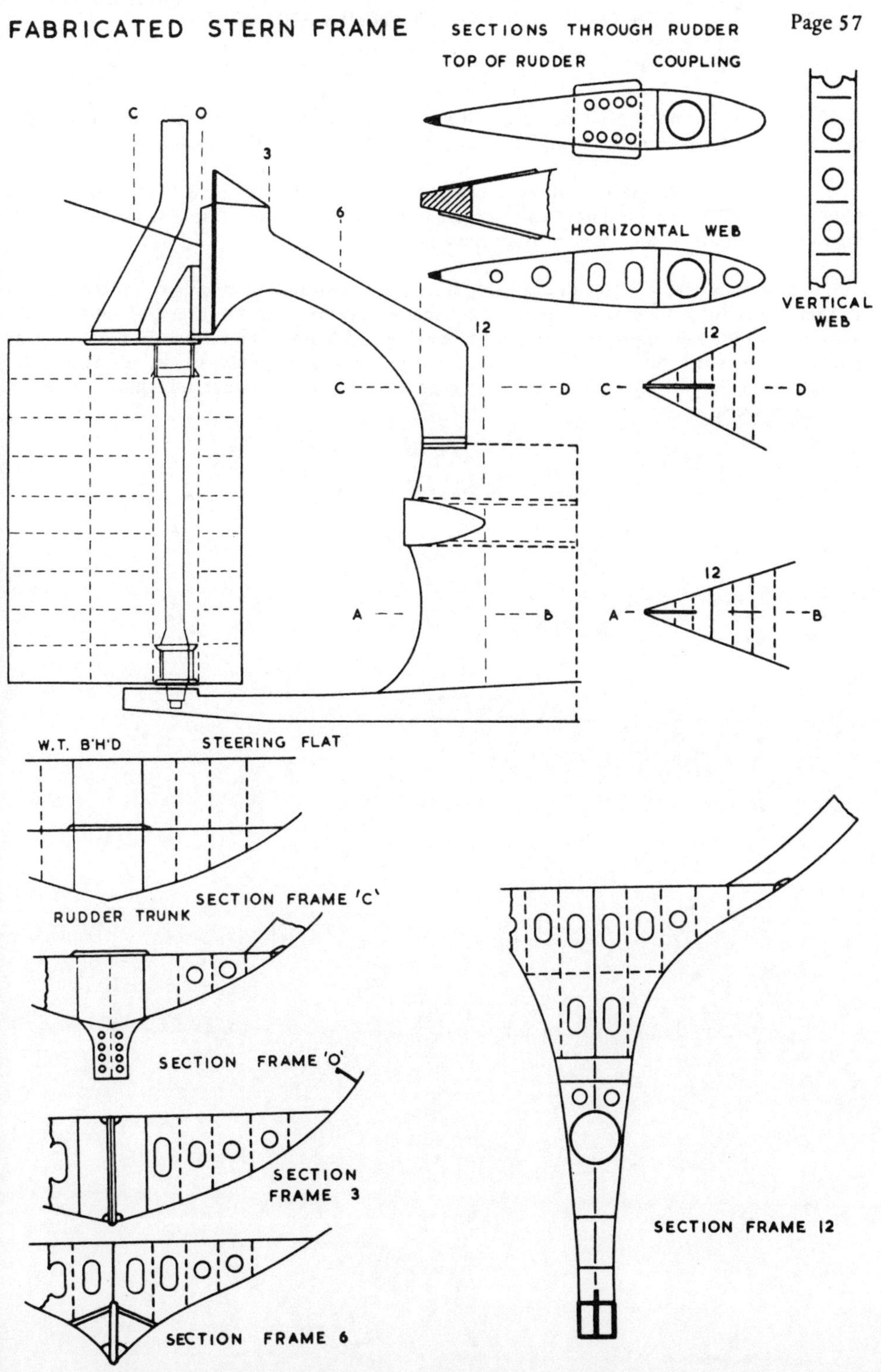

The stern frames of twin or quadruple screw vessels do not have to support the tail shaft and propeller, their only function being to support the rudder.

The various illustrations show a cast stern frame and a fabricated stern frame for a twin screw vessel. In the latter case sections in way of the stern frame at various frames forward and abaft of the transom floor have been illustrated.

In a twin screw vessel the tail shafts may be enclosed by bossings, and supported at their ends by a spectacle frame, or the shafts may be exposed after leaving the hull, with the after ends supported by an 'A' bracket or frame. Occasionally both may be found in the same vessel, a short portion of the shafts being enclosed by a bossing and the remainder of the shafts exposed and supported by an 'A' frame, but this is not usual on large vessels.

FABRICATED STERN FRAME TWIN SCREW

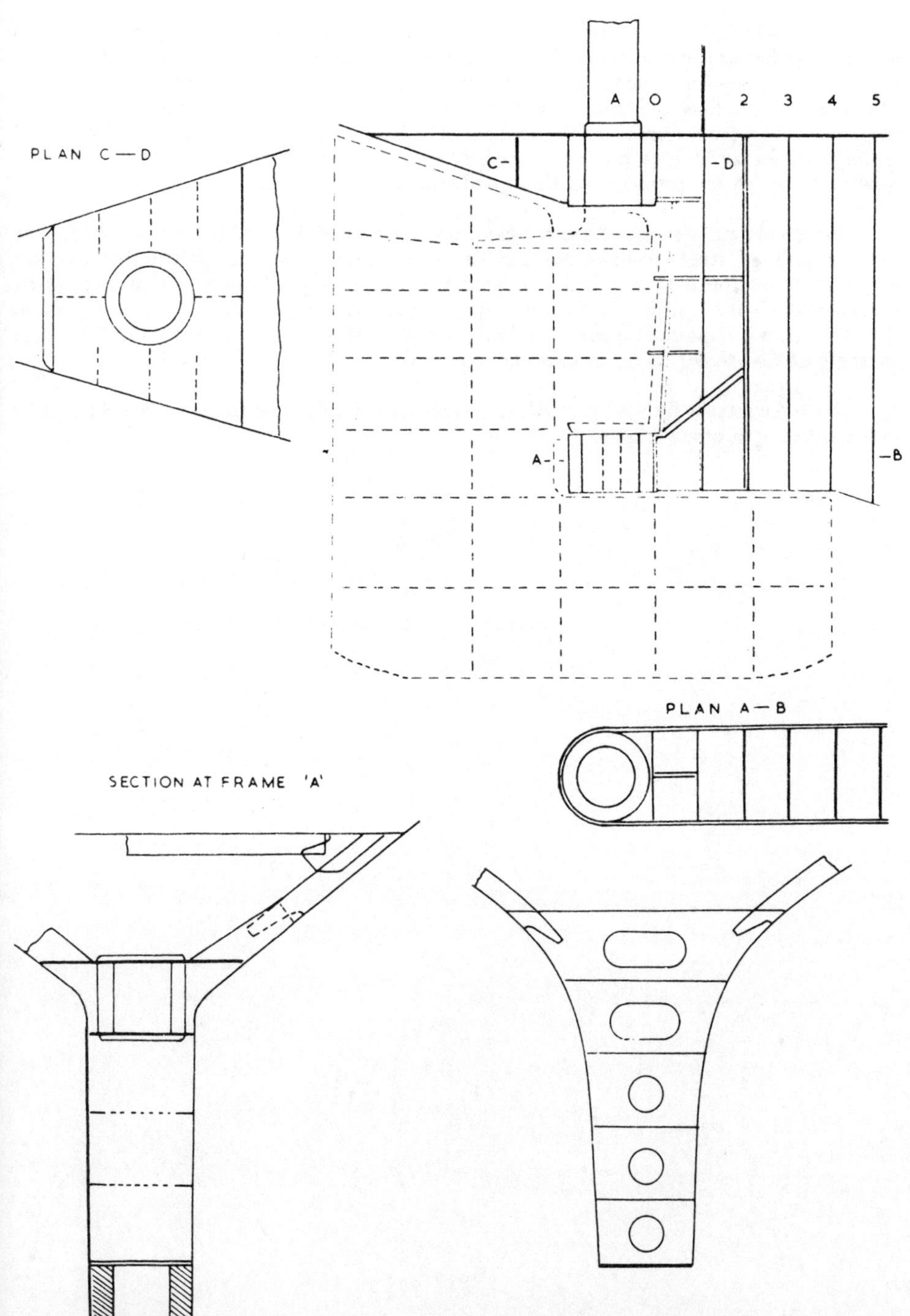

A 'Spectacle Frame' may consist of two castings attached to the stern section and welded together at the centre line, or in a very large vessel extend only far enough inboard of the shell plating for an adequate connection to be obtained with the adjacent structure. The illustrations show the initial curvature of the shell plating at the after end to that of the actual 'spectacle frame'. The bossing as will be seen is continuous with the shell plating and is faired to a fine trailing edge abaft the spectacle frame so as to allow the flow of water to the propeller to be as undisturbed as possible.

Where plated bossings or a spectacle frame are not used an 'A' bracket is fitted. The two streamlined struts forming the bracket will usually pass through the shell and the inboard connection is then made to a system of brackets and frames so that stresses are transmitted to the adjacent structure. Watertightness is maintained by welding round the strut where it passes through the hull. In order that an 'A' bracket should be rigid an angle of 60° to 90° is usual between the struts.

Both 'spectacle frames', and 'A' frames are used though the former is more frequent in the larger twin screw vessels.

PROPELLER FRAMES TWIN SCREW VESSELS

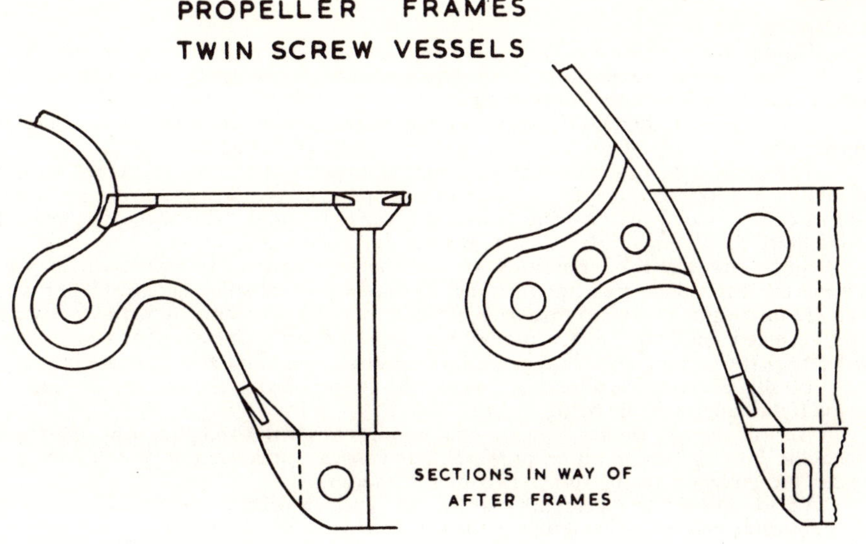

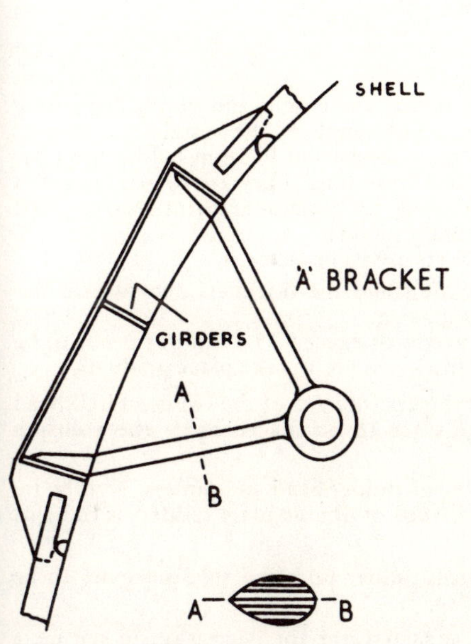

'A' BRACKET

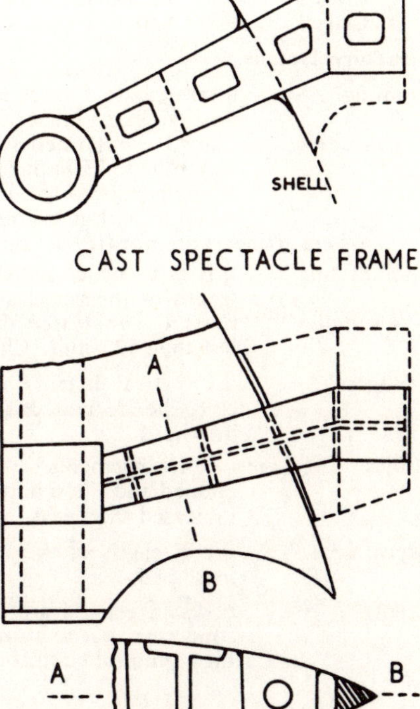

CAST SPECTACLE FRAME

RUDDERS

Double plate rudders may be balanced or unbalanced depending on the size of the vessel. The several illustrations, in association with stern frames, show various types of rudder, their construction and bearings.

The shape of the rudder plays an important part in its efficiency. The area is approximately 2% of the product of length and designed draught.

The construction of a double plate rudder consists of two plates welded on to a cast or forged steel frame and stiffened with horizontal and vertical webs or it may be entirely fabricated of welded plates. The rudder stock may be solid, tubular or of a special shape to facilitate the fitting of plates. See separate illustrations.

Since the vertical dimensions of a rudder are necessarily restricted, in order to achieve the equivalent area, the fore and aft dimensions must be increased. The resultant increase in torque needed to turn such a rudder may be overcome by fitting a balanced or semi-balanced rudder. Such a rudder has approximately one-third of the rudder area forward of the turning axis. A balanced or semi-balanced rudder is illustrated.

Ideally a balanced rudder is one where the centre of pressure and the turning axis coincide for all angles of the helm.

An unbalanced rudder consists of a number of pintles and gudgeons, the top pintle being the locking pintle which prevents any vertical movement in the rudder and the pintle and gudgeon taking the weight of the rudder.

Various other arrangements may be met with in practice depending on the design of the rudder and whether designed for single or twin screw vessels.

Strengthening for Navigation in Ice.

The vertical extent of ice strengthening is related to the Light and Load Waterlines. Four classes of ice strengthening are detailed in the Rules, these are :—

Class 1* — extreme ice conditions
Class 1 — severe ice conditions
Class 2 — intermediate ice conditions
Class 3 — light ice conditions

Structural Details.

Framing: — intermediate frames are to be fitted, the extent and depth depending on the Class. Generally over the full length of the vessel.

Example: Class 1 — intermediate frames to extend from 915mm below the light waterline to 750mm above the load waterline. They are to be connected at their ends to the adjacent framing by horizontal members or carried down to within 255mm of the margin plate.

Where stringers are not fitted, tripping brackets are to be fitted.

Shell Plating: — this is to be of increased thickness and the thickness depends on the length of the vessel and frame spacing.

Example: Class 1 — increased thickness need not exceed 25.5mm but is not to be less than 12.5mm. Changes in thickness are to take place gradually.

Stringers — the vertical distance apart of stringers forward of the collision bulkhead is to be decreased and stringers are to be spaced abaft the collision bulkhead.

Rudder and Steering Arrangements: the diameter of rudder head and pintles is to be increased and the side plating and webs of double plate rudders is to be of increased thickness.

Sternframe — the strength of the rudder horn, rudder post and sole piece are to be increased.

Stem: — bows in general are to be of special form for navigation in ice i.e. a solid stem bar to above the load waterline. Where plate stems are fitted they should be reinforced by a centre line web and by horizontal webs.

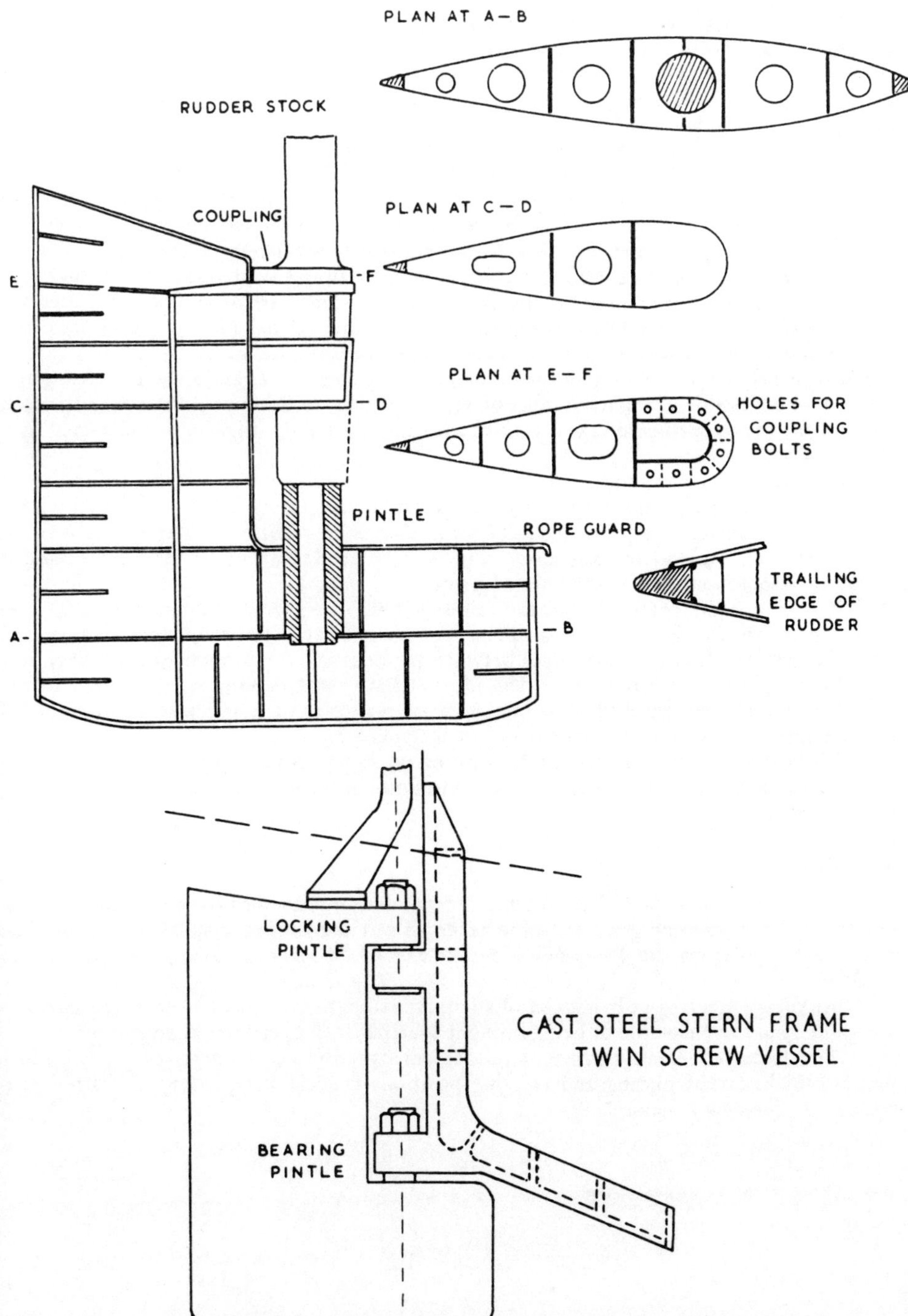

CAST STEEL STERN FRAME TWIN SCREW VESSEL

STERN TUBES

Stern tubes are fitted to provide a bearing for the Tail End Shaft and to enable a watertight gland to be fitted at an accessible position. The tube is usually constructed of cast steel with a flange at its forward end and a thread at the after end. It is inserted from forward and this end is bolted over packing to the after peak bulkhead. A large nut is placed over the thread at the after end, tightened and secured to the propeller post.

A bronze liner is fitted to the tail end shaft, usually for the full length of the stern tube and a watertight gland is fitted at the forward end. The after end of the tube is fitted with a brass bush, and strips of lignum vitae are set into this, providing the necessary bearing for the shaft. Lubrication is by sea water which is free to enter the after end of the tube. Laminated plastics may be used as a substitute for lignum vitae.

Where vessels operate frequently in water containing sand or sediment, the wear down on the lignum vitae may be excessive and stern tubes are often fitted in which sea water is excluded and the bearings are lubricated by oil. Bronze liners need not be fitted to tail end shafts in such cases since the bearings will be of white metal. Where this system is used it is necessary to fit a gland at each end of the stern tube, and since the after gland will not be accessible in service it must be self adjusting. In the type illustrated a flange is attached to the propeller so that it rotates with the shaft and oiltightness is obtained by a rotating gland.

The after end of the tail end shaft is tapered to receive the propeller boss and a key is provided to transfer the torque from the shaft to the propeller. A nut, fitted with a locking plate, secures the propeller in position and as an additional safeguard it is fitted with a left hand thread in association with a right handed propeller, or a right handed thread in association with a left handed propeller.

To remove the propeller and the tail end shaft, the propeller should be slung, after removing the rope guards, and the propeller nut slacked back. The propeller is then started from the shaft by driving steel wedges between the boss and the propeller post. When it is free the propeller nut is removed. The intermediate shaft (length of shaft next to the tail end shaft) is next removed. The tail end shaft may then be withdrawn into the tunnel and the propeller can then be removed from the aperture.

Withdrawal of the tail end shaft is necessary every three years (four years for ships with two or more screws) if fitted with continuous liners or oil glands, in all other cases every two years.

STEERING GEAR

All ships are to be provided with two independent means of moving the rudder. With power operated steering gears the time taken to put the rudder from 35 degrees on one side to 30 degrees on the other side is not to exceed 28 seconds at the maximum service speed.

Auxiliary steering gear is to be of adequate strength, sufficient to steer the ship at a navigable speed and capable of being brought speedily into operation in an emergency.

The Rules cover all aspects of rudder construction i.e. diameter of stock, rudder pintles, thickness of plating and steering positions especially regarding the latter with respect to Passenger Vessels.

TAIL END SHAFT AND STERN TUBE

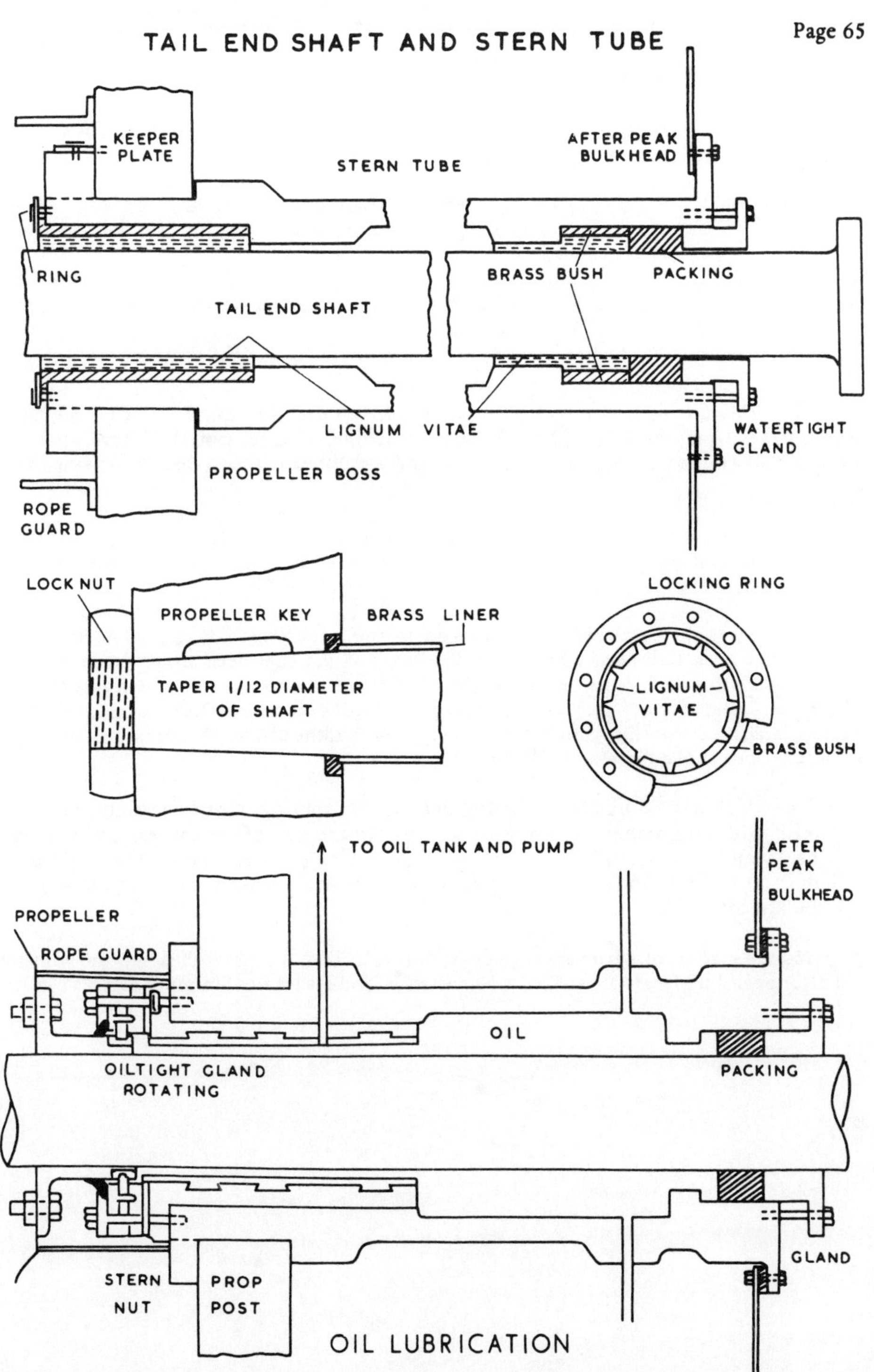

OIL LUBRICATION

WATERTIGHT BULKHEADS

Transverse watertight bulkheads which divide a vessel into a number of watertight compartments are of great importance for the following reasons :—

(1) Strength: they give large structural support, resist any tendency to deformation (racking) and assist in spreading the hull stresses over a large area.

(2) Fire: confines conflagration to particular regions.

(3) Subdivision: divides a vessel into a number of watertight compartments.

All ships are to have a collision bulkhead, situated not less than 0.05 L nor more than 0.075L from the fore end of the load waterline, an after peak bulkhead enclosing the stern tubes in a watertight compartment and a bulkhead at each end of the machinery space.

Additional watertight bulkheads are to be fitted depending on the length of the vessel. Structural compensations are to be made where the number of bulkheads is below Rule number.

All watertight bulkheads are to extend to the uppermost continuous deck, except for the after peak bulkhead which may terminate at the first deck above the load waterline provided this deck is made watertight to the stern or to a watertight transom floor. Where the draught is not greater than that permitted with a superstructure extending for the full length of the ship above the second deck, bulkheads may terminate at that deck provided it lies above the load waterline.

The thickness of bulkhead plating depends on spacing of stiffeners and height of bulkhead, with a minimum thickness of 5.5mm. In the case of ore carriers the minimum is to be 10mm. The scantlings of stiffeners are to be calculated, the section modulus of stiffeners on collision bulkheads being 25% greater than that required for ordinary watertight bulkheads.

No set spacing of stiffeners is given in the Rules but in general those on the collision and after peak bulkheads are spaced 610mm apart and elsewhere 760mm apart.

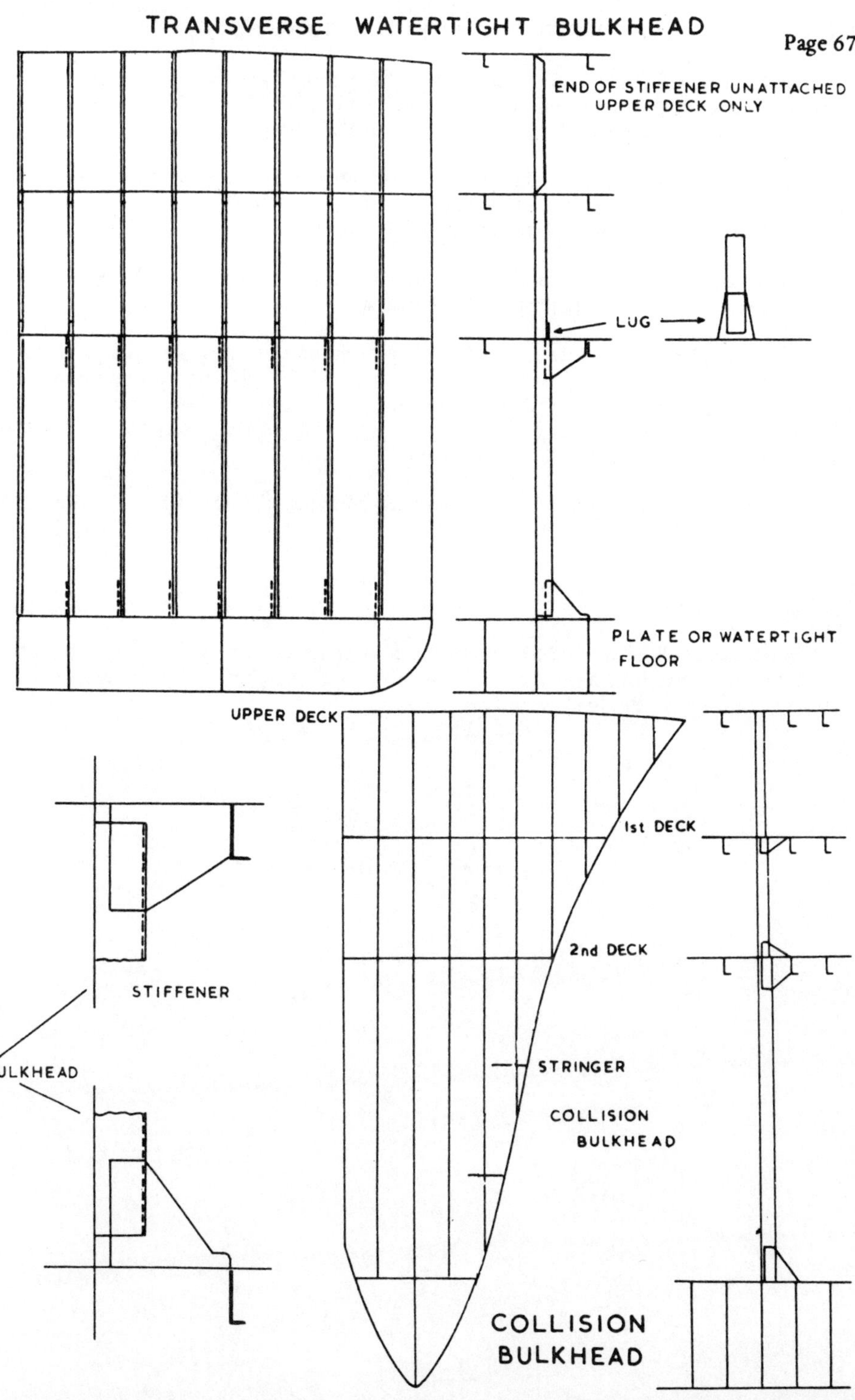
TRANSVERSE WATERTIGHT BULKHEAD
END OF STIFFENER UNATTACHED UPPER DECK ONLY
LUG
PLATE OR WATERTIGHT FLOOR
UPPER DECK
1st DECK
2nd DECK
STIFFENER
BULKHEAD
STRINGER
COLLISION BULKHEAD
COLLISION BULKHEAD

In Passenger Vessels many other factors have to be taken into account when considering the number and spacing of watertight bulkheads. The reader is referred to the Merchant Shipping Passenger Ship Construction Rules 1965.

If there are openings in watertight bulkheads, watertight doors with suitable framing must be fitted, and additional stiffening in way of the doors must be fitted so that strength is the same as that of an unpierced bulkhead.

Pipes and valves attached direct to the bulkhead plating are to be secured by studs screwed through the plating or by welding.

Corrugated bulkheads are now frequently found in tankers, general cargo ships, ore carriers etc. They are usually trapezoidal in shape and though easy to prefabricate they require heavy presses to give the required shape. They afford a considerable reduction in welding, less welding results in reduced buckling, and are less susceptible to corrosion (more efficient tank cleaning).

Superstructure bulkheads are occasionally swedges, the spacing of swedged being 610mm or less.

Testing

Watertight bulkheads including recesses and flats are to be hose tested on completion. Peak bulkheads not forming boundaries of tanks are to be tested by filling peaks with water to the level of the load waterline.

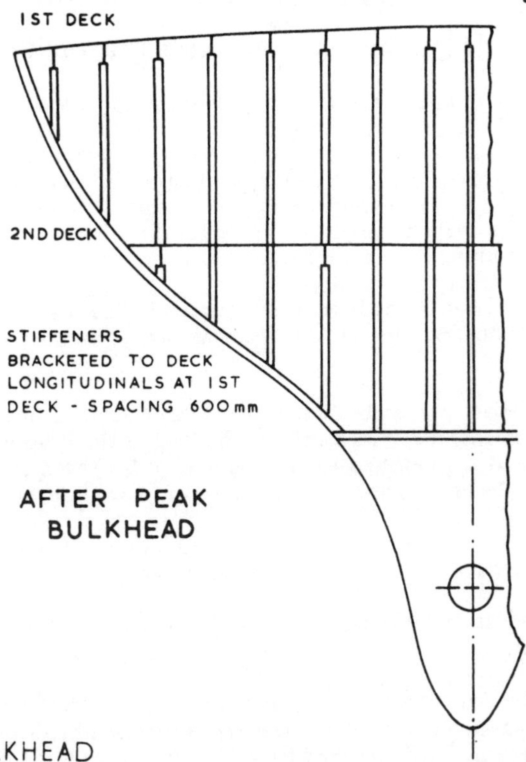

AFTER PEAK BULKHEAD

CORRUGATED BULKHEAD

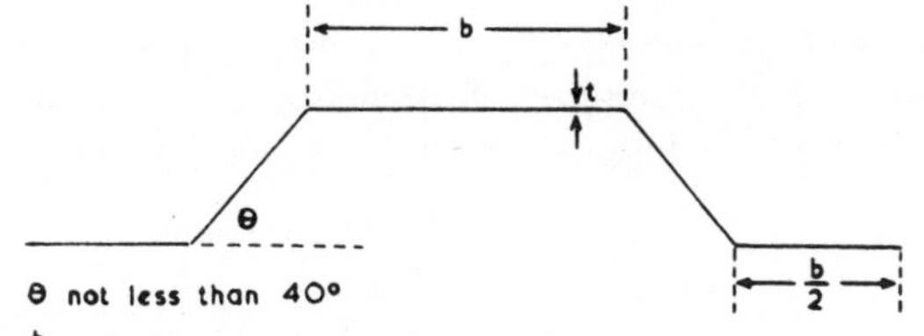

θ not less than 40°

$\frac{b}{t}$ not to exceed 70 at bottom
85 at top of bulkhead

SWEDGE

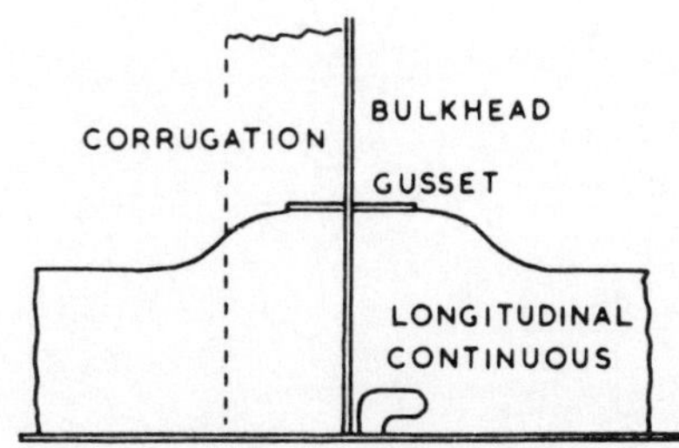

ARRANGEMENT OF BOTTOM LONGITUDINAL IN WAY OF O.T. TRANSVERSE BULKHEAD

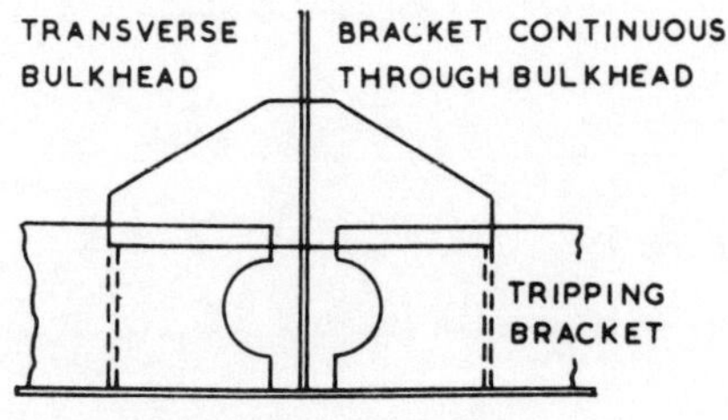

LONGITUDINAL WITH ENDS SCALLOPED

PILLARS, GIRDERS AND NON-WATERTIGHT BULKHEADS

The primary function of pillars and girders is to transmit the deck loads to the bottom structure where the distributed loads are supported by the upthrust of the water pressure. They also tie the vessel together vertically thus preventing the flexing of the decks in response to the bending of the side frames under varying water pressures and vertical accelerations in heavy weather. Normally a pillar will be in compression though in certain cases it is possible for the pillar to be subjected to tension and even side loads resulting from the movement of cargo when rolling.

In cargo holds large widely spaced hollow tubular pillars are usual. These tend to reduce 'broken stowage' and such an arrangement is frequently referred to as 'Massed Pillaring'.

Pillars are to be fitted in the same vertical line wherever possible and effective arrangements made to distribute the load at the head and heel. Pillars are to be securely bracketed at their heads with doubling plates fitted at the head and heel, which should be over the intersection of a plate floor and side girder without manholes under the pillar.

Where pillars are not arranged directly above the intersection of floors and girders then partial floors and intercostals are to be fitted.

Pillars to be fitted below deck houses, windlasses, winches, etc. to give the necessary support.

In lieu of pillars non-watertight pillar bulkheads may be fitted on the centre line. They usually extend from the transverse bulkhead to the hatch coaming – such an arrangement facilitates the fitting of shifting boards when carrying grain cargoes.

The thickness of the plating is not to be less than 7.5mm in holds and 6.5mm in tween decks. Stiffeners are not to be more than 1.5m apart. Stiffeners or corrugations are to have a depth of at least 150mm in holds and 100mm in tween decks. Pillar bulkheads may be swedged.

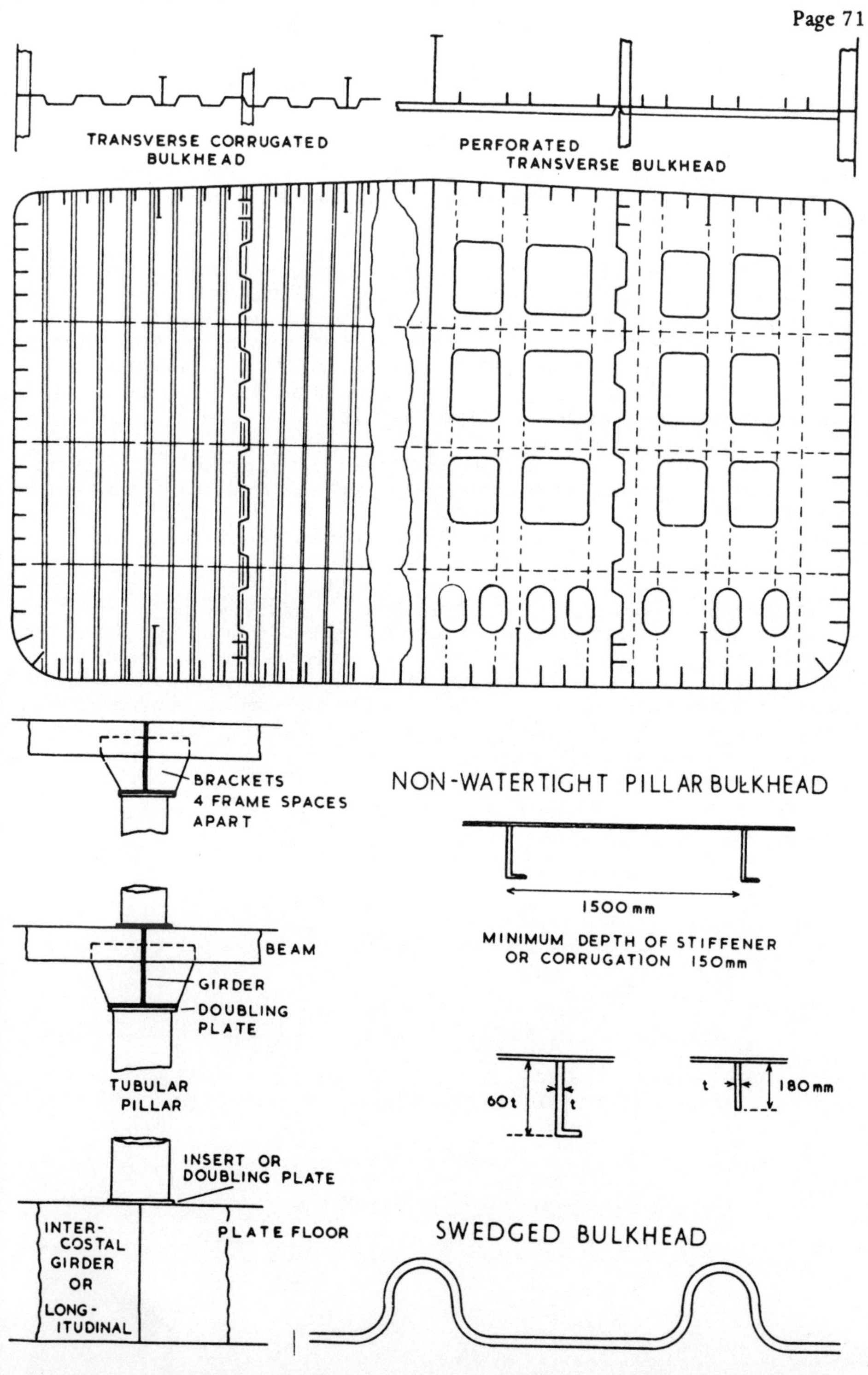
TRANSVERSE CORRUGATED
BULKHEAD
PERFORATED
TRANSVERSE BULKHEAD
BRACKETS
4 FRAME SPACES
APART
NON-WATERTIGHT PILLAR BULKHEAD
1500 mm
MINIMUM DEPTH OF STIFFENER
OR CORRUGATION 150mm
BEAM
GIRDER
DOUBLING
PLATE
TUBULAR
PILLAR
60t
t
t
180mm
INSERT OR
DOUBLING PLATE
INTER-
COSTAL
GIRDER
OR
LONG-
ITUDINAL
PLATE FLOOR
SWEDGED BULKHEAD

HATCHWAYS

Hatchways in the majority of dry cargo ships extend across the deck for approximately one-third of the beam. In special types of vessels, i.e. containerships, colliers, etc., much wider hatchways are fitted as will be seen from the separate cross sections of these types.

Special arrangements must be made to compensate for the structural discontinuity caused by these large openings; inset plates of increased thickness may be required at the hatch corners as shown. The arrangement of hatch coaming and adjacent structure with rounded hatch corners is illustrated. Note that the hatch coaming should be extended beyond the corner to form a bracket, see Deck Girders. The deck openings should be well rounded off to a reasonable radius, preferably eliptical or parabolic in shape, to avoid a concentration of stress at these points, see page 34.

The deck plating forms an important structural member, especially at the strength deck, in resisting longitudinal stress, but only the strakes clear of the hatches can be considered in this respect. If hatchways are made unduly wide, the effective width of this plating is reduced and the thickness will have to be increased in order that the cross sectional area of the steel shall be maintained.

In addition to the plating beams will also be cut at hatchways and the ends of the half beams will be connected to the hatch coamings and supported by deck girders. The deck girders are usually integral with the hatch coamings as illustrated. At the ends of the hatchways, in the case of transverse framing and at the sides in the case of longitudinal framing deep hatch end beams will be fitted to support the coamings. A pillar will often be placed near the hatch corners at the intersection of the deck girder and strong beam.

HATCH CORNERS

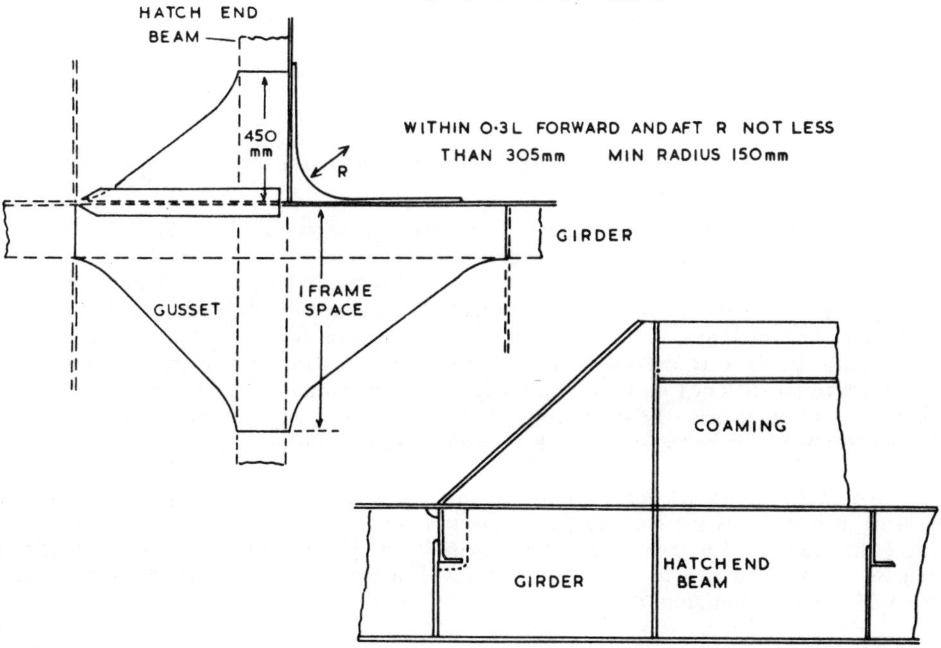

HATCH END
BEAM
450
mm
WITHIN 0·3L FORWARD AND AFT R NOT LESS
THAN 305mm MIN RADIUS 150mm
R
GIRDER
GUSSET
1 FRAME
SPACE
COAMING
GIRDER
HATCH END
BEAM

l1 not less than 2l2
l2 = b/20
b
b = breadth hatchway
opening
not less than 305 mm nor greater than 610mm

One frame space fore and aft
760mm
760mm
min. radius within
0.6L amid 305mm
forward and abaft this
region 150 mm

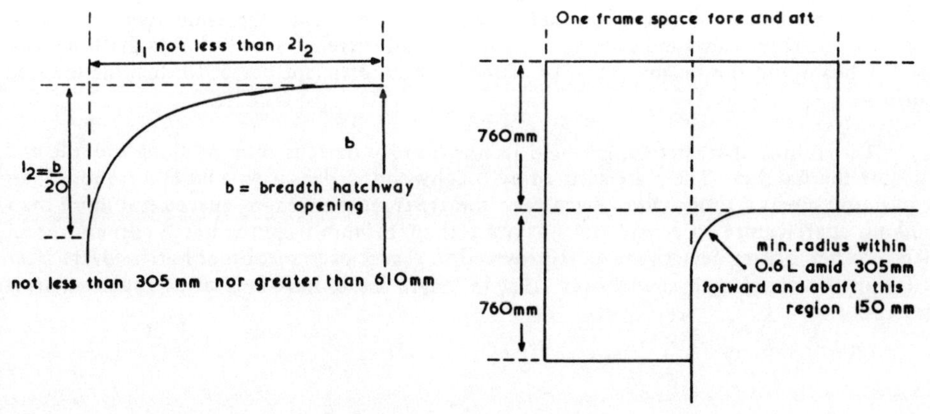

Hatch coamings on weather decks are not to be less than 600mm in height in Position 1 and 450mm in height in Position 2. To meet Factory Act requirements they are usually 760mm.

Position of Hatchways.

Position 1 – Hatchways on exposed freeboard decks, and exposed superstructure decks within forward 0.25L.

Position 2 – Hatchways on exposed superstructure decks abaft 0.25L.

Hatch coamings are not to be less than 11mm thick and coamings 600mm or more in height are to be stiffened on their upper edge by a horizontal bulb flat or equivalent not less than 180mm in depth. Additional support is to be given by brackets or stays from the bulb flat to the deck at intervals of not more than 3m. Coamings less than 600mm in height are to be stiffened on their upper edge by a substantial moulding. Side coamings are to extend to at least the lower edge of the deck beams and the lower edge is to be flanged to avoid damage to cargo and the fraying or runner wires.

Where hatch covers are used in conjunction with portable beams and indirect securing arrangements the wooden hatch covers will have a thickness not less than 60mm where the span is 1.5m and 82mm with a span of 2m, proportionate thickness for intermediate spans. The ends of the hatch covers will be protected by galvanised steel bands about 65mm wide and 3mm thick.

The hatch covers rest on beams as illustrated. The hatch beams consist of webs stiffened at their upper and lower edges. The ends of the webs are to be doubled or inserts fitted at least 180mm wide. Those beams which carry the ends of the hatch covers (king beams) are to be fitted with a 50mm vertical flat on the upper surface and the bearing surface for the covers is to be not less than 65mm. The portable beams are supported at their ends by carriers or sockets having a minimum bearing surface of 75mm.

Where roller or sliding beams are fitted they conform to the same standards as for ordinary portable beams and carriers. The horizontal stiffener usually forms the trackway for the beam and the beams are to be fitted with an efficient device for locking them in position.

Tarpaulins, at least two, are held in position by battens running along the side and ends of the hatches. These are secured by toughwood wedges with a taper not more than 1 in 6 and cleats 65mm wide, set to take the taper of the wedges, spaced not more than 600mm apart centre to centre and not more than 150mm from the hatch corners. Steel bars or other locking devices are to be provided to secure each section of hatch covers after battening down. Hatch covers over 1.5m in length are to have at least two such securing devices.

HATCHWAYS

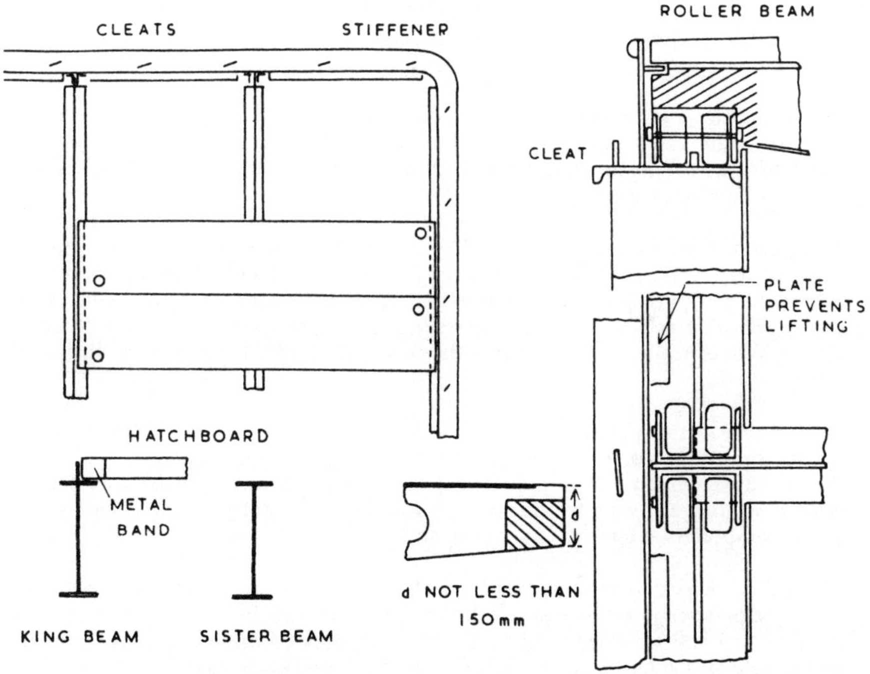

CLEATS
STIFFENER
ROLLER BEAM
CLEAT
PLATE PREVENTS LIFTING
HATCHBOARD
METAL BAND
d
d NOT LESS THAN 150mm
KING BEAM
SISTER BEAM

180mm
BRACKET OR STAY AT INTERVALS OF 3m
COAMING OVER 600mm IN HEIGHT

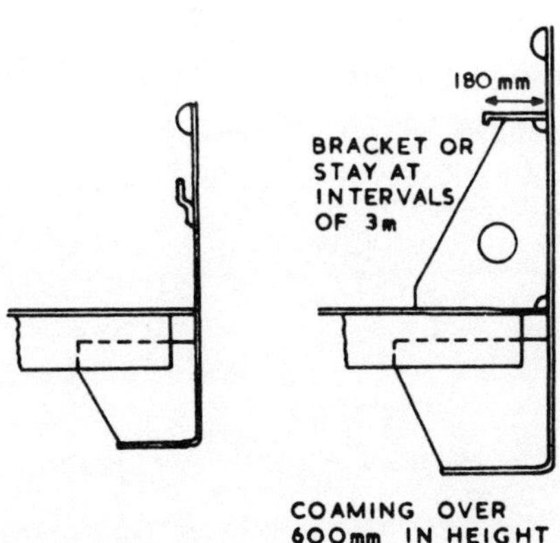

Steel Hatch Covers having Direct Securing Arrangements, i.e. Macgregor Steel Hatches are a great improvement on the type previously mentioned and are now standard for weather decks. They consist of plated covers stiffened by webs or stiffeners, watertightness being obtained by gaskets and clamping devices.

Securing cleats and cross joint wedges, together with suitable jointing material are to be fitted, the cleats being not more than 2m apart with a minimum of 2 per panel at the sides and with one arranged adjacent to each corner at the hatch ends. The cross joint wedges are to be spaced about 1.5, apart.

The portable sections are connected to one another and can easily and quickly be rolled into or out of position leaving clear hatchways and decks. The normal practice is for the lengthwise opening of hatches but sideways opening hatches are found on some particular types of vessels, i.e. OBO carriers, see pages 79 and 119. They may be operated manually or hydraulically. The illustrations show the simple type of manually operated hatch cover.

The wheels at the ends of the hatch sections, eccentric rollers, are used for raising the section clear of the coaming and for rolling it along the coaming trackway. As shown the axles of these wheels are so adjusted that when the hatch is in the closed position the weight is no longer borne by them. The jointing fits tightly on the coaming and the hatch made completely watertight by fitting and securing the cleats.

The roller is used when the hatch cover is pulled in to its stowage position. It engages on the plate edge at the ends of the hatchway and enables it to be turned into the vertical. Wires, chains or bars attached to the stub axles of these rollers at the centre of the wheel enable all the hatch covers to be drawn back and forth together.

The cross joints are made watertight as shown with cross joint wedges.

There are a large number of different types of steel hatches each serving a similar purpose.

Details of gastightness with reference to OBO carriers are illustrated on page 79.

STEEL HATCHES

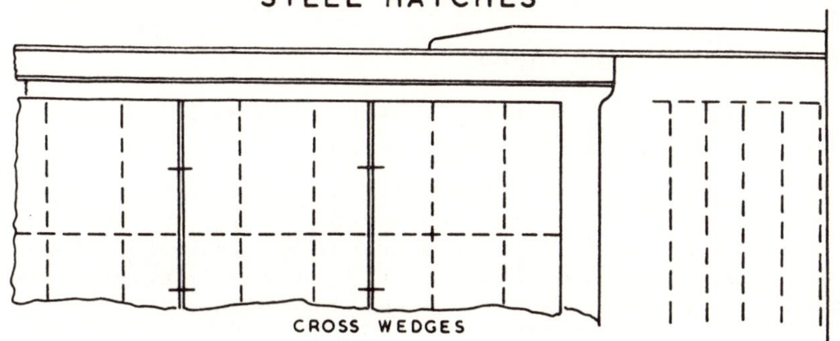
CROSS WEDGES

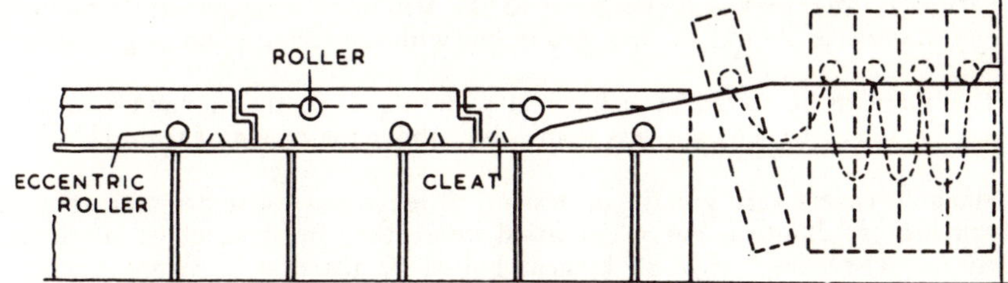
DIRECT SECURING ARRANGEMENTS
ROLLER
ECCENTRIC ROLLER
CLEAT

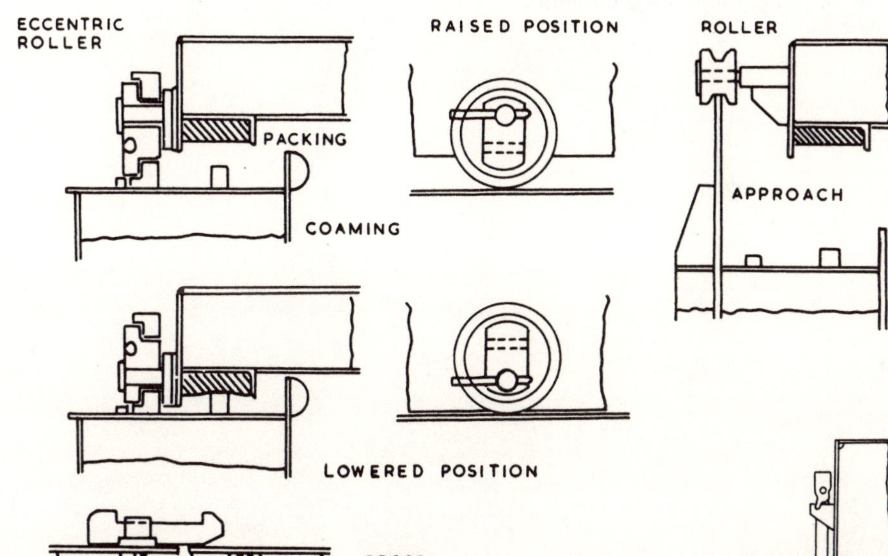
ECCENTRIC ROLLER
RAISED POSITION
ROLLER
PACKING
COAMING
APPROACH
LOWERED POSITION
CROSS WEDGE
QUICK ACTING CLEAT

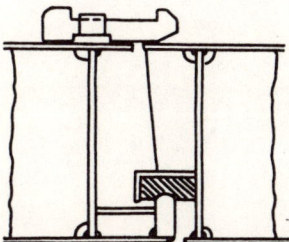

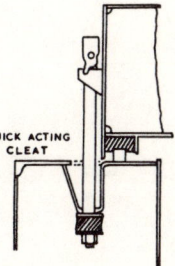

PEAK, DEEP AND TOPSIDE TANKS

The closing arrangements for deep tanks are as illustrated. Peak and topside tanks are usually closed by manhole covers

A centre line bulkhead is to be fitted in deep tanks which are designed for the carriage of oil fuel bunkers and which extend the full breadth of the vessel. This bulkhead may be intact or perforated. If intact, the construction is similar to that for an ordinary transverse watertight bulkhead. If the bulkhead is perforated the area of perforation is to be between 5 and 10 percent of the total area of the bulkhead.

If girders are fitted in the tank they are to form a continuous line of support on the bulkheads and ship's sides.

Particular attention is to be given to the structural arrangements within topside tanks. A transverse should be arranged in line with the end of main cargo hatchways.

Tanks are to be tested by a head of water equal to the maximum to which the tank may be subjected, but not less than 2.44m above the crown of the tank.

Manhole covers vary greatly in design. When fitted outside a tank they may be either circular or elliptical but when fitted inside they must be elliptical to facilitate withdrawal. The covers may be hinged, bolted or fitted with strong backs. Usual size of opening 450mm by 600mm.

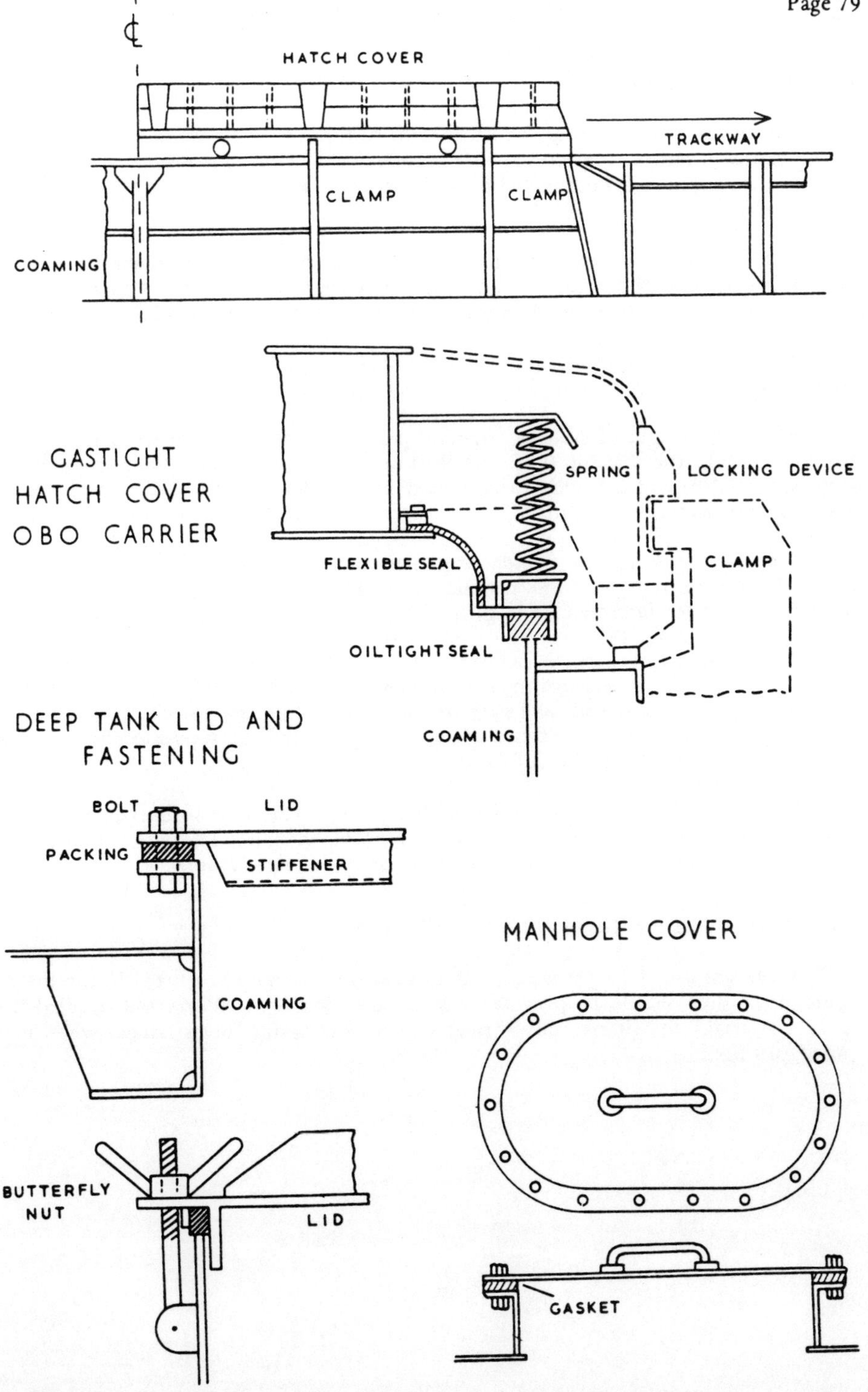

HATCH COVER
TRACKWAY
CLAMP
CLAMP
COAMING
GASTIGHT
HATCH COVER
OBO CARRIER
SPRING
LOCKING DEVICE
CLAMP
FLEXIBLE SEAL
OILTIGHT SEAL
COAMING
DEEP TANK LID AND
FASTENING
BOLT
LID
PACKING
STIFFENER
COAMING
MANHOLE COVER
BUTTERFLY
NUT
LID
GASKET

SUPERSTRUCTURES

All seagoing vessels are to be fitted with forecastles or increased sheer. The height of the exposed deck at the side (at fore perpendicular) above the summer load waterline is to be not less than that determined from the Rules – this is a function of the length and block coefficient.

Forecastles are to extend from the stem to a point at least 0.07L abaft the forward perpendicular or if the required bow height is obtained by means of increased sheer, the sheer is to extend for at least 0.15L from the forward perpendicular.

The thickness of fore ends, sides and after ends of all superstructures and deckhouses are to be of or over a minimum thickness of 6.5mm.

Where the length of a bridge superstructure exceeds 0.15L or part of an end superstructure within the midship 0.5L, the thickness of the upper deck sheerstrake is to be increased by 50% and the deck stringer plate by 25% above Rule thickness at the ends. Bridge superstructures less than 0.15L, the increases are to be 30% and 10% respectively.

The standard height of a bridge superstructure depends on the length of the vessel and is 1.8 metres if L is 75 metres and 2.3 metres of L is 125 metres or more.. Intermediate heights for intermediate lengths of L.

As mentioned on pages 34 and 36, attention is to be given to the structure at the corners and ends, and the side plating of superstructures having a length of 0.15L or greater is to be increased at the ends and is to be tapered into the upper deck sheerstrake. This plating is to be efficiently stiffened along its upper edge and supported by web plates not more than 1.5m from the bulkhead.

Where large deckhouses are fitted above superstructures web frames or partial bulkheads are to be fitted within the superstructure 9m apart. They are also to be fitted in way of large openings, boat davits and other points of high loading.

PORT HOLES, SIDE LIGHTS OR SIDE SCUTTLES

These are usually constructed with a brass frame, and when fitted below the freeboard deck they should be provided with a steel deadlight. Ports and deadlights are made watertight by fitting rubber rings which bear against brass flanges when in the closed position.

SUPERSTRUCTURE BULKHEAD

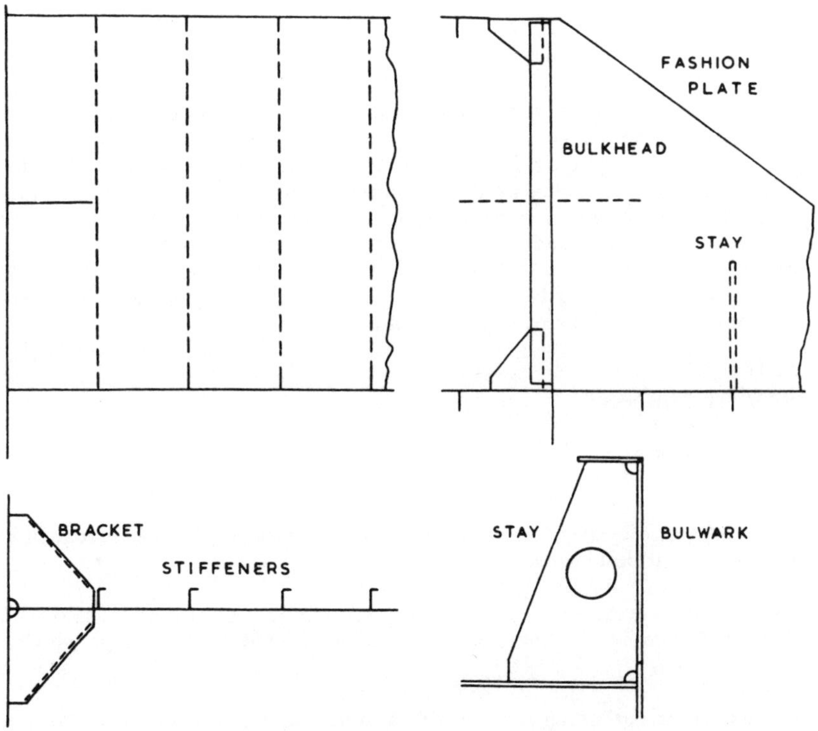

FASHION PLATE

BULKHEAD

STAY

BRACKET

STIFFENERS

STAY

BULWARK

PORT HOLE

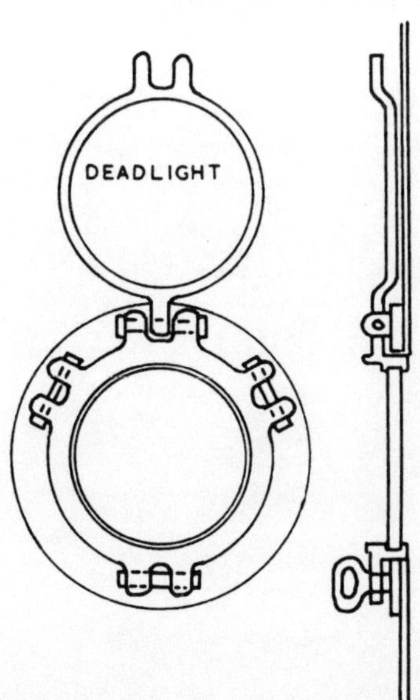

WATERTIGHT DOORS

Both the hinged and sliding type are shown illustrated. The sliding type is frequently operated by hydraulic rams in place of the hand gear. The Rules state that watertight doors are to be efficiently constructed and fitted, tried under working conditions, capable of operating with a list of 15 degrees, and finally hose tested when in place. They are to be capable of being quickly closed from an accessible position above the bulkhead deck and are to have an index at the operating position showing whether the door is open or closed. Hand gear is to have an all round motion. The lead of the shafting is to be as direct as possible and the screw is to work in a gunmetal nut. Hinged watertight doors of approved pattern may be fitted in between decks in approved positions. The hinges of these doors are to be fitted with gunmetal pins. They are to be secured in the closed position by means of dogs or clamps.

WATERTIGHT TUNNELS

A sliding watertight door, capable of being operated on both sides of the door, is to be provided at the forward end of the tunnel.

The thickness of plating and scantlings of stiffeners is to be calculated in a similar manner to watertight bulkheads.

Under hatchways the top plating is to be increased by 2mm unless covered with wood not less than 50mm in thickness which is to be secured by fastenings which do not penetrate the plating.

Additional strengthening is to be fitted under the heels of pillars or masts stepped on the tunnel.

Tunnels are to be hose tested on completion.

WATERTIGHT DOORS

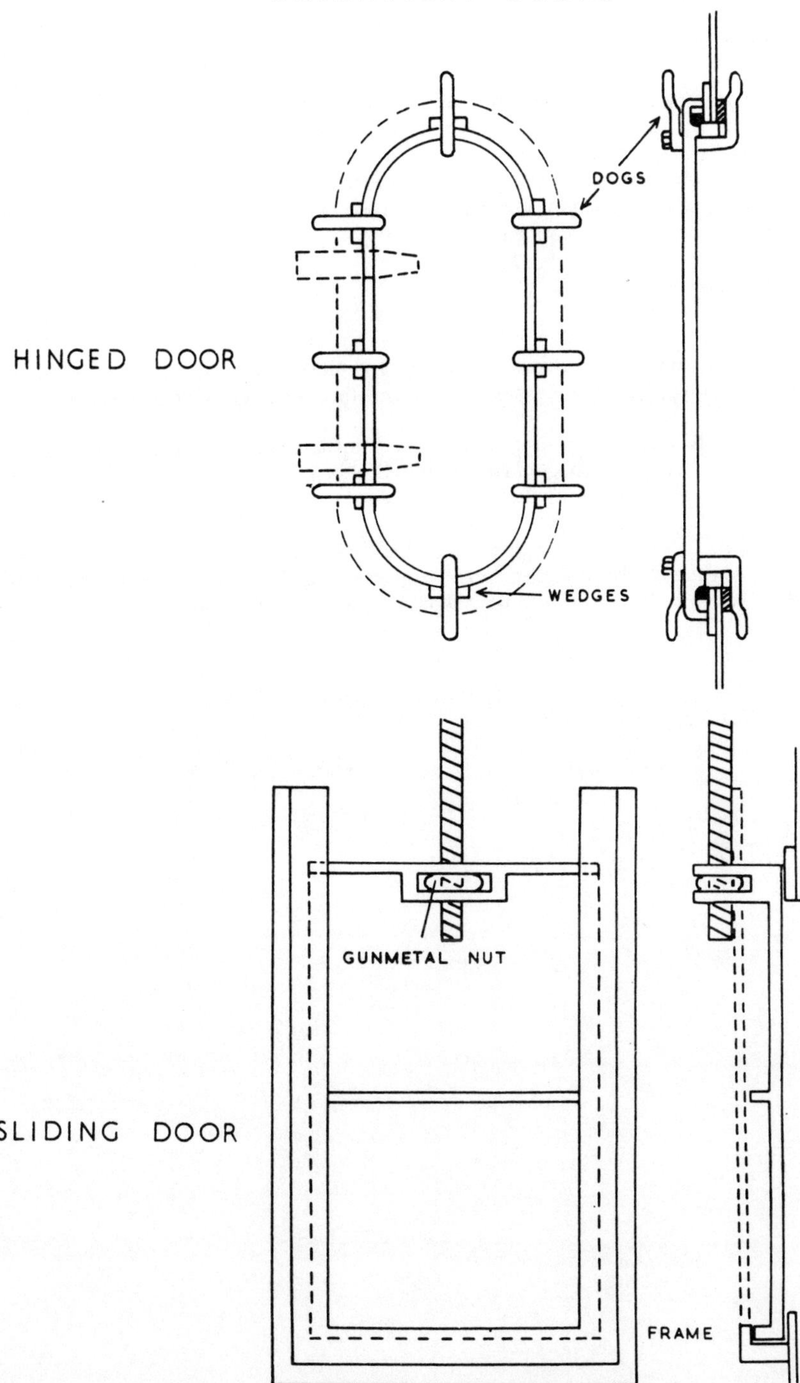

CARGO DOORS

Cargo doors are fitted in certain trades to provide access to between deck spaces, i.e. direct loading by fork lift truck from the quay into the .tween deck.

Openings are cut into the shell plating and arrangements must be made to maintain the strength, particularly in a longitudinal direction. The corners of all openings should be well rounded to avoid stress concentrations.

Illustrated are :–

(1) a cargo port, manually operated, secured by closely spaced dogs or bolts. This arrangement is typical of the type fitted to facilitate the loading of stores, etc.

(2) a hydraulically operated sliding door shown in the open and closed positions. This type is simple and fast to operate and is self closing since the door is forced against the perimeter of the opening due to the eccentric path of its guide rollers.

(3) a swing door. This type of door may be fitted at the sides of the vessel to give access to 'tween deck or at the stern to give access for vehicles i.e. Ro Ro Vessels. In the latter case the ramps will be a separate piece of equipment.

CARGO DOORS

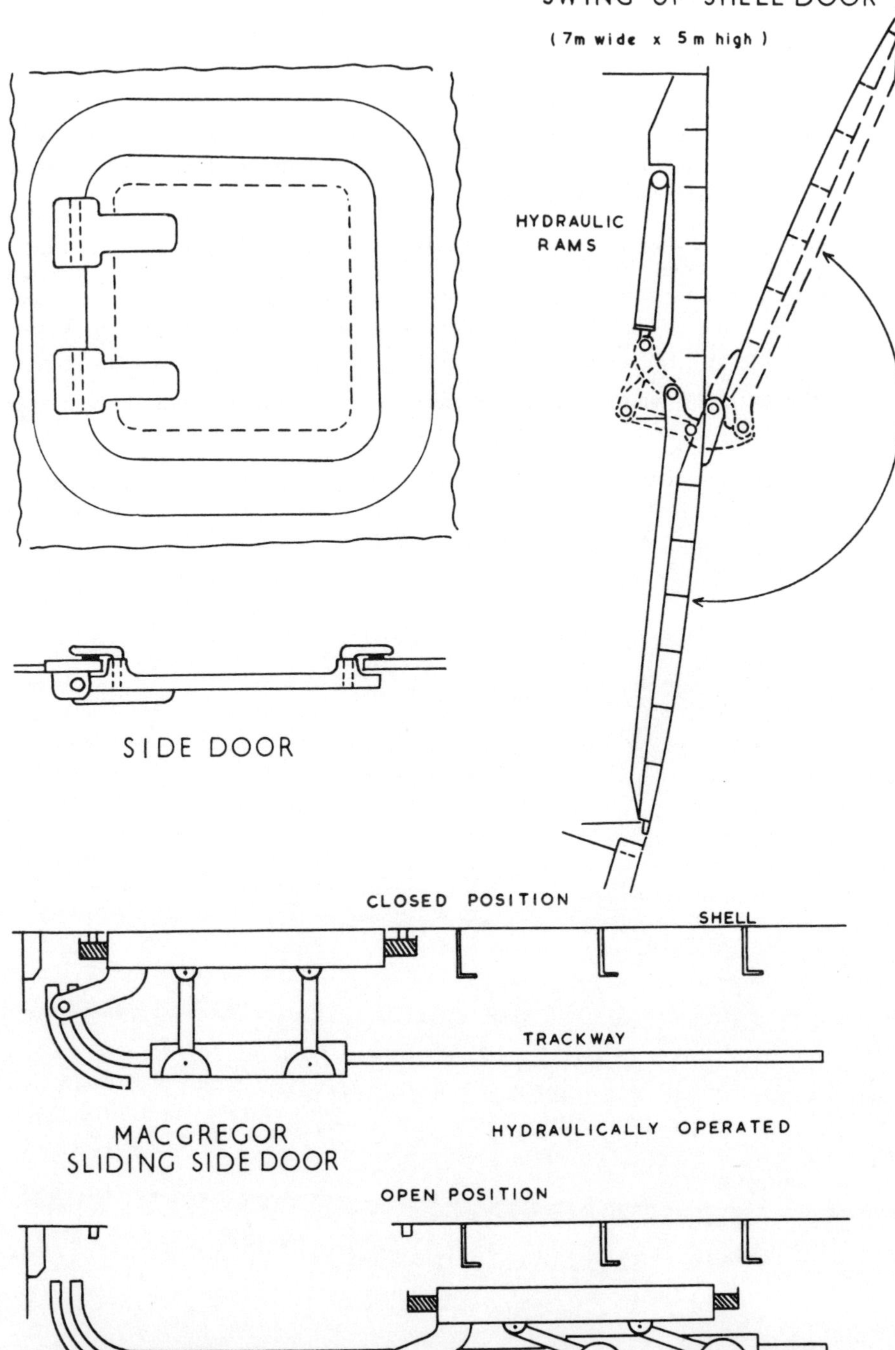
SWING UP SHELL DOOR
(7m wide x 5m high)
HYDRAULIC
RAMS
SIDE DOOR
CLOSED POSITION
SHELL
TRACKWAY
MACGREGOR
SLIDING SIDE DOOR
HYDRAULICALLY OPERATED
OPEN POSITION

BOW AND STERN DOORS AND RAMPS

The illustrations show the arrangements at the bow and stern of roll on/roll off vessels.

The vessel illustrated on page 129 permits of through movement of vehicles. Many roll on/roll of vessels only have stern doors and ramps. The arrangements vary depending on the nature of the service being operated.

When using the bow doors it is necessary for the bow of the vessel to run into a specially designed fender. Watertight closure at the bow is provided by either the ramp or fairing piece i.e. bow door. Where watertightness is provided by the bow ramp in its raised position the fairing piece has a spray tight seal. A watertight bow door is illustrated.

At the stern the ramp when in the raised position usually forms the watertight closure as shown.

The ramps and doors are normally hydraulically operated and cleated in position.

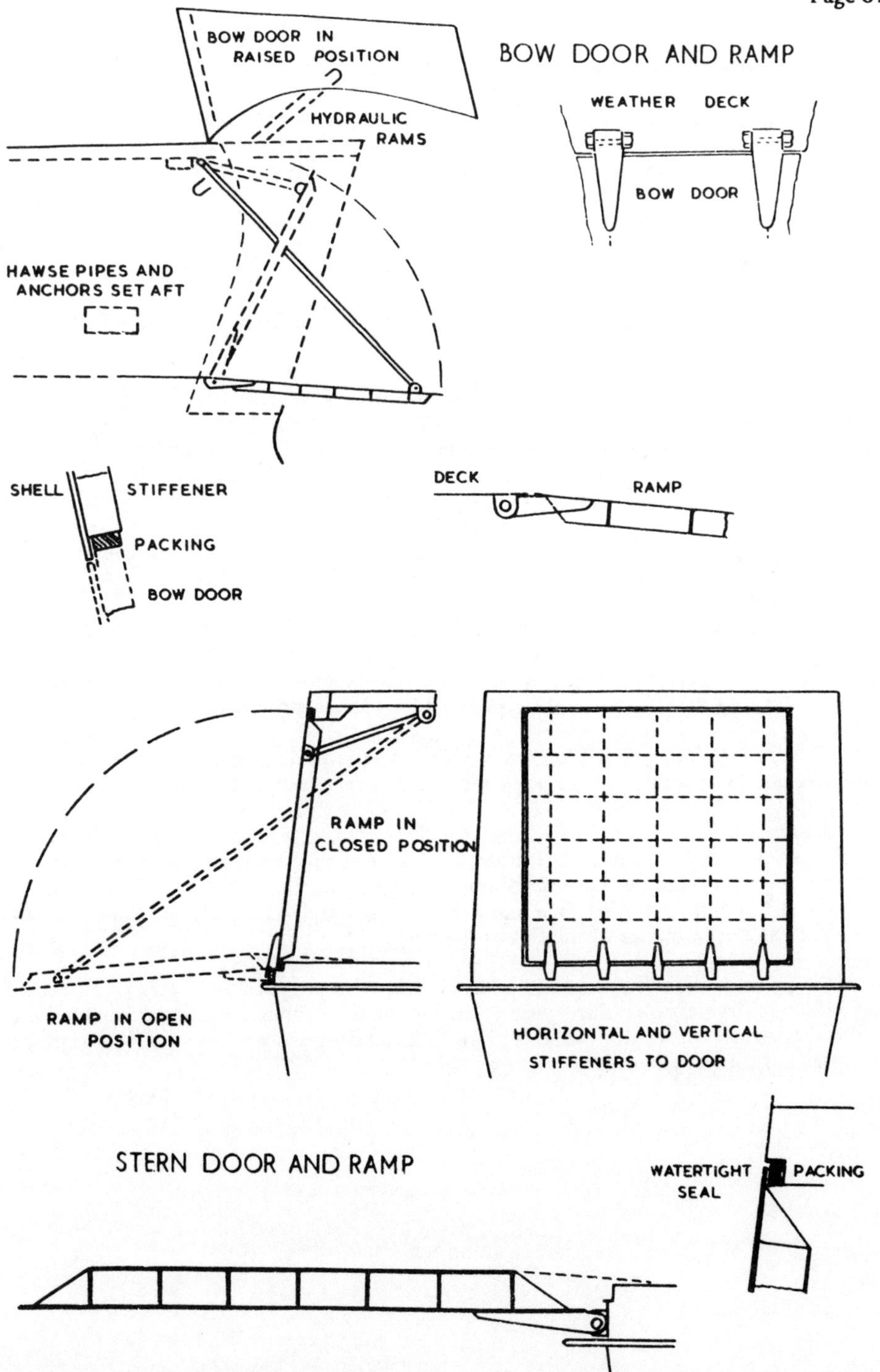
BOW DOOR AND RAMP
BOW DOOR IN RAISED POSITION
HYDRAULIC RAMS
HAWSE PIPES AND ANCHORS SET AFT
WEATHER DECK
BOW DOOR
SHELL
STIFFENER
PACKING
BOW DOOR
DECK
RAMP
RAMP IN CLOSED POSITION
RAMP IN OPEN POSITION
HORIZONTAL AND VERTICAL STIFFENERS TO DOOR
STERN DOOR AND RAMP
WATERTIGHT SEAL
PACKING

BULWARKS AND GUARD RAILS

Bulwarks and guard rails are fitted for the safety of the crew and play no part in the structural strength of the vessel. They are usually 1 metre in height. plated bulwarks are to be stiffened by a strong rail section and supported by stays from the deck.

Spacing of stays to forecastle bulwarks is not to exceed 1.2m on A, B-60 and B-100 ships (see Load Line Rules) and not more than 1.83m on other types. Elsewhere the stays are not to be fitted more than 1.83m apart.

The plating of bulwarks is to be doubled or of increased thickness in way of mooring pipes, eye plates, etc.

Where bulwarks form wells ample provision is to be made for freeing the decks of water as rapidly as possible. Bulwarks must therefore be provided with Freeing Ports, the area on each side of the vessel is to be calculated and depends on the length of the well.

The lower edge of the freeing point is to be as near to the deck as possible and the openings are to be protected by rails spaced approximately 230mm apart. If hinged doors or shutters are fitted to freeing ports they must have ample clearance to prevent jamming and the pins or bearings are to be of non-corrodible material.

If the construction of bulwarks was made integral with the sheerstrake then the light plating of the bulwarks (being further from the neutral axis) would be subjected to considerable stress with the possibility of subsequent fracture. This would create a notch at the sheerstrake and might give rise to a serious structural fracture.

A 'floating' bulwark is illustrated which is suitable for welded construction. Bulwarks are not to be welded to the top edge of the sheerstrake within 0.5L amidships.

Where guard rails are fitted they shall consist of courses of rails supported by stanchions efficiently secured to the deck. The opening between the lowest course of rails and the deck shall not exceed 230mm in height and above that course openings shall not exceed 380mm in height. Where the ship has a rounded gunwale the stanchions shall be secured at the perimeter of the flat of the deck.

Scuppers sufficient in number and size are to be provided and fitted in all decks to give effective drainage. In ships over 150m in length scupper openings are not to be cut in the sheerstrake above deck level within 0.5L amidships nor in way of discontinuities, i.e. breaks of superstructures.

BULWARKS

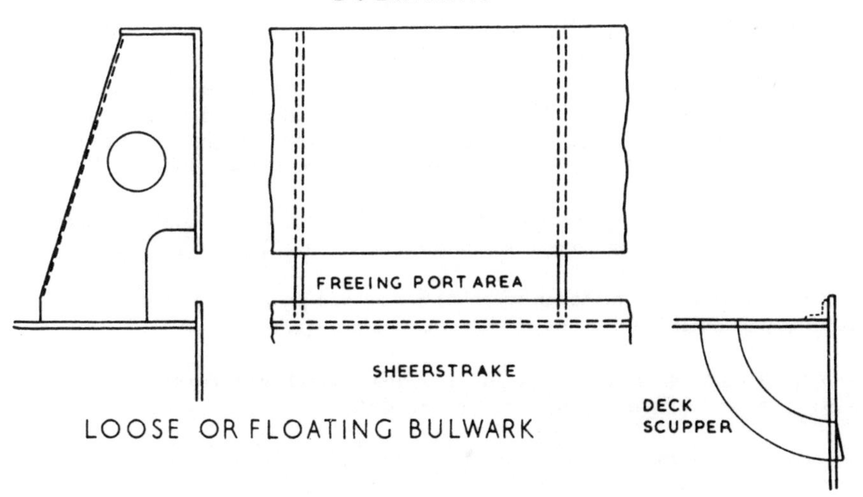
FREEING PORT AREA
SHEERSTRAKE
LOOSE OR FLOATING BULWARK
DECK
SCUPPER

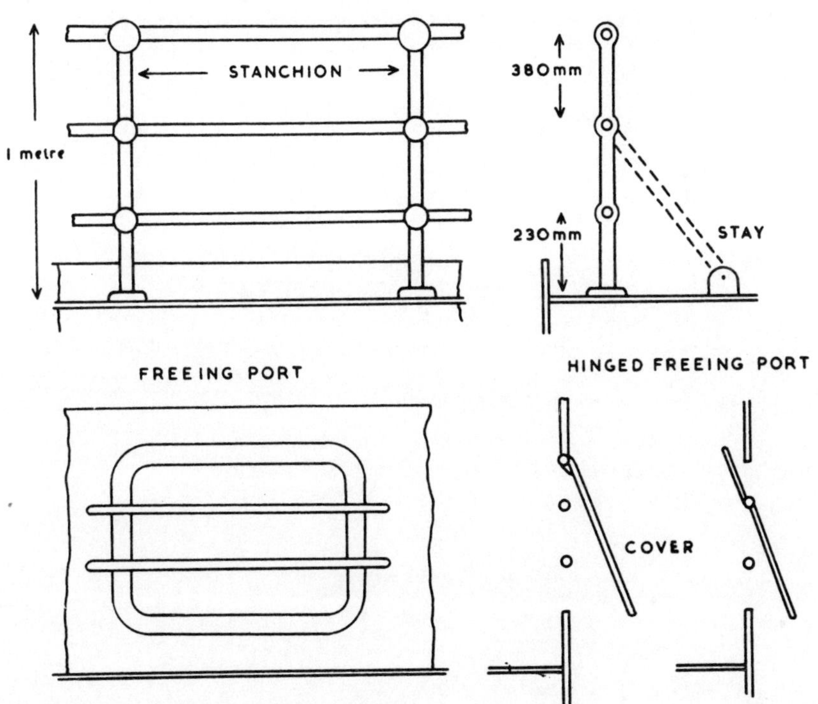
RAILS AND FREEING PORTS
STANCHION
1 metre
380mm
230mm
STAY
FREEING PORT
HINGED FREEING PORT
COVER

VENTILATORS, AIR AND SOUNDING PIPES

Ventilators are necessary to give adequate air circulation to under deck spaces, accommodation and tanks. The coamings are to have a minimum height above the surface of weather decks of 900mm in Position 1 and 760mm in Position 2 (see page 74). Where coamings exceed 900mm in height they are to specially stayed.

All ventilator coamings are to be supplied with strong plugs and canvas covers unless the height of the coaming exceeds 4.5m in Position 1 and 2.3m in Position 2.

Special care is to be taken when designing and positioning ventilator openings and coamings particularly in regions of high stress concentration.

Mushroom, gooseneck and other similar minor ventilators are to be strongly constructed and efficiently secured to the deck.

Goose or swan neck type ventilators are mainly used for the air pipes to tanks. The height (H) as shown is not to be less than 760mm on the freeboard deck and 450mm on the superstructure deck. Air pipes are to be fitted at the opposite end of the tank to that which the filling pipe is placed and/or at the highest point of the tank.

Sounding pipes are to be as straight as possible and to have a bore not less than 32mm. Where a sounding pipe passes through a refrigerated compartment where temperatures may be 0°C or below the bore is not to be less than 65mm. Striking plates of adequate thickness and size are to be fitted under open ended sounding pipes.

VENTILATORS

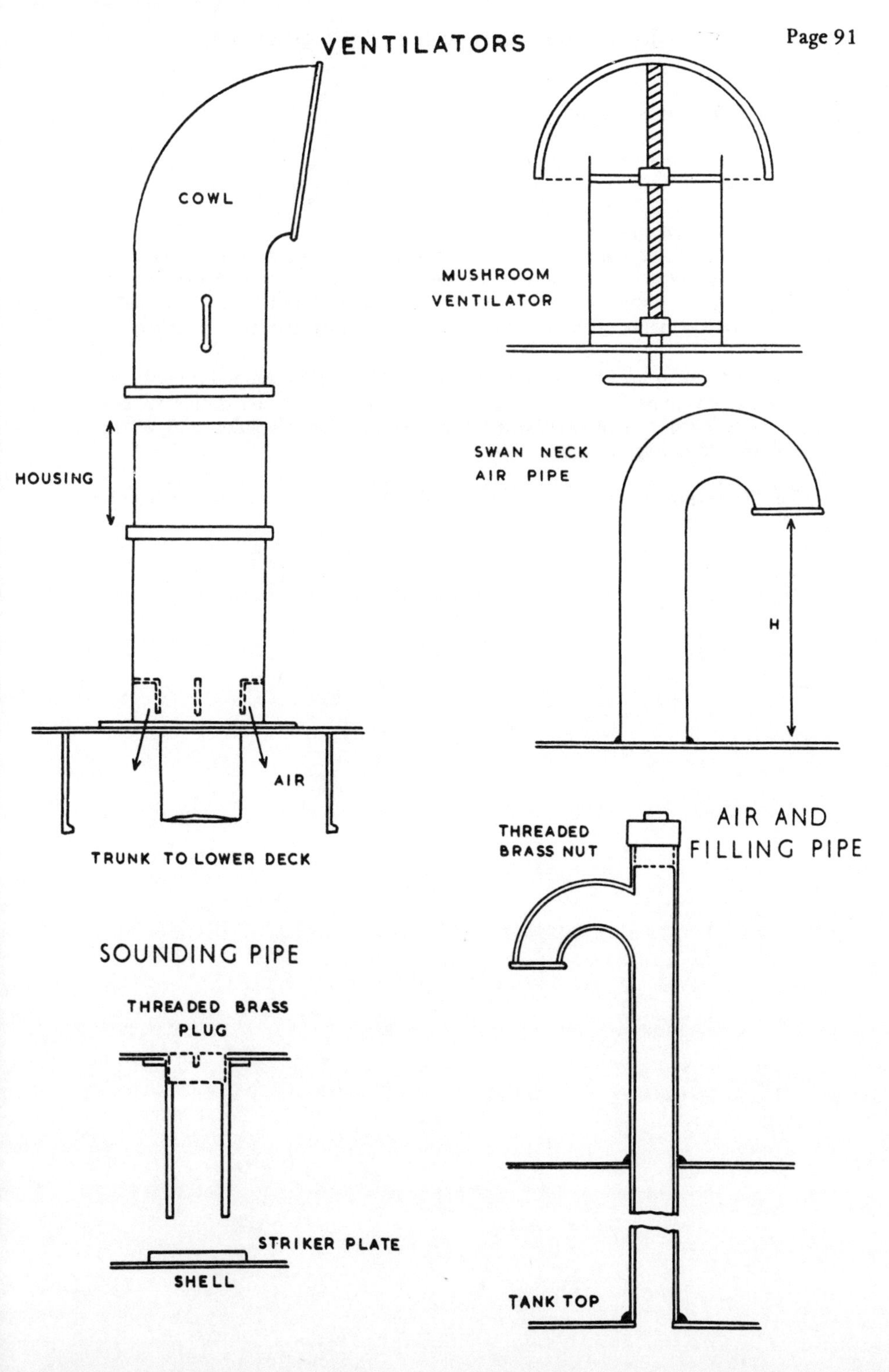
COWL
HOUSING
AIR
TRUNK TO LOWER DECK
MUSHROOM
VENTILATOR
SWAN NECK
AIR PIPE
H
AIR AND
FILLING PIPE
THREADED
BRASS NUT
TANK TOP
SOUNDING PIPE
THREADED BRASS
PLUG
STRIKER PLATE
SHELL

MAST AND DERRICK POSTS

These are subjected to stresses when the derricks are lifting weights and their minimum thickness will depend on their length and safe working load.

Stresses are greatest at the points of attachment so that the plating is of increased thickness at the heel, deck and in way of derrick fittings. The deck plating will be doubled or as is usual with welded construction, an insert of increased thickness will be fitted at the heel with adequate support from below. They should normally be attached to at least two decks and the heels of masts and derricks should be effectively supported.

The accompanying illustrations show a mast (or pillar) stepped directly on a tunnel, and a mast of the type fitted on board tankers. In the first case notice the large bracket fitted directly beneath the heel which transfers the thrust of the mast to the adjacent bulkhead and adjoining structure.

The Rules now give mast scantlings and rigging sizes, based on derricks being at an angle of 30 degrees to the horizontal.

MAST STEPS

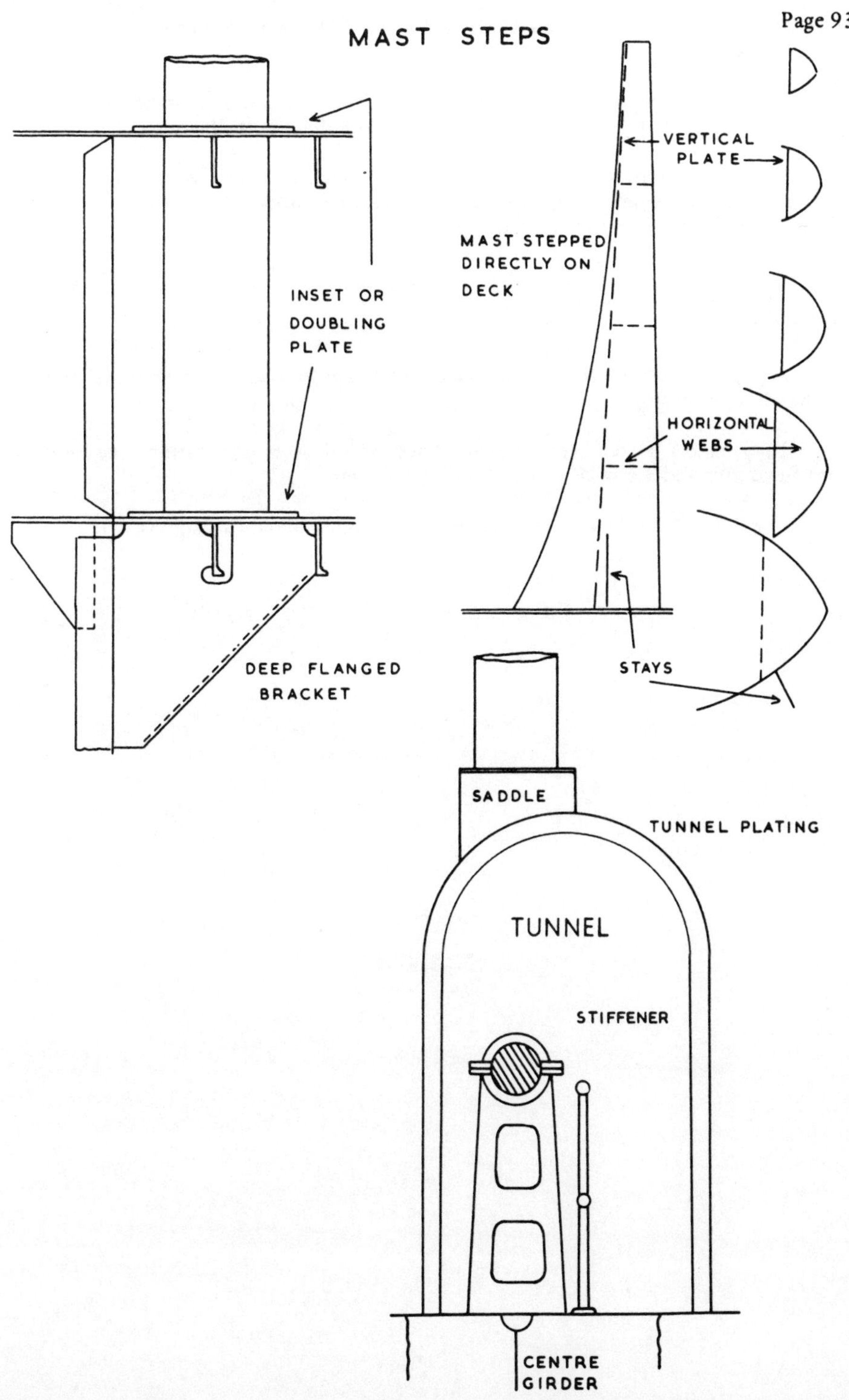

VERTICAL PLATE
MAST STEPPED DIRECTLY ON DECK
INSET OR DOUBLING PLATE
HORIZONTAL WEBS
DEEP FLANGED BRACKET
STAYS
SADDLE
TUNNEL PLATING
TUNNEL
STIFFENER
CENTRE GIRDER

Many ships are now fitted with Hallen, Stulcken and similar heavy derricks and lifting gear. An outline arrangement of a Stulcken derrick is illustrated.

Derrick goosenecks, illustrated, allow a derrick to be swung in a horizontal, vertical or oblique plane. Means must be provided to prevent the pins from lifting and, as shown, a check nut is fitted for this purpose.

DECK CRANES

A large number of vessels are fitted with cranes instead of conventional derricks for the handling of cargo.

They have a better overall performance, 360 degrees of rotation, one hundred per cent flexibility and are faster in the handling of cargo.

The illustration shows a typical deck crane in use aboard a cargo vessel.

OUTLINE STULCKEN DERRICK

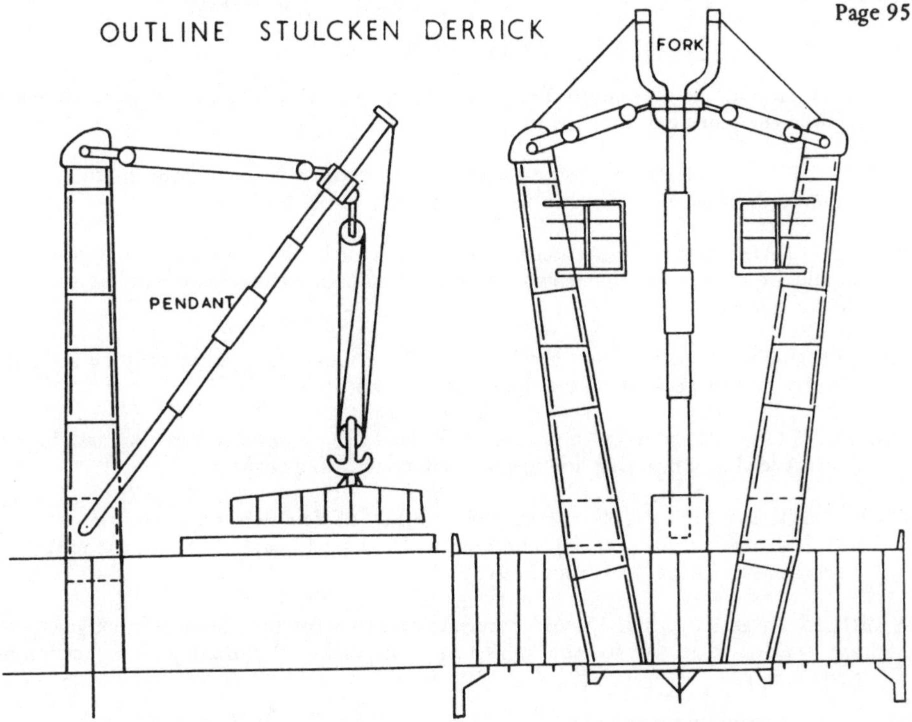

SIDE ELEVATION FRONT ELEVATION

DECK CRANE

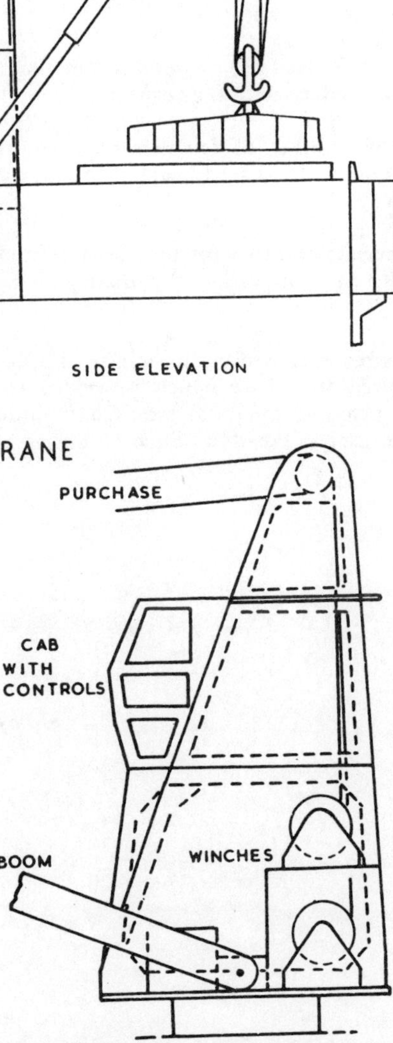

ROTATING BASE

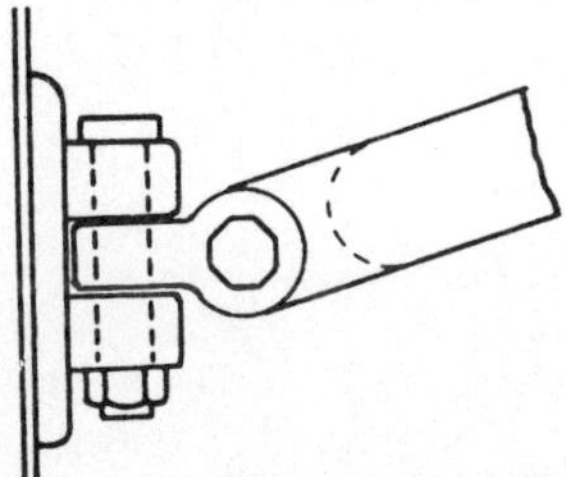

DERRICK GOOSENECK

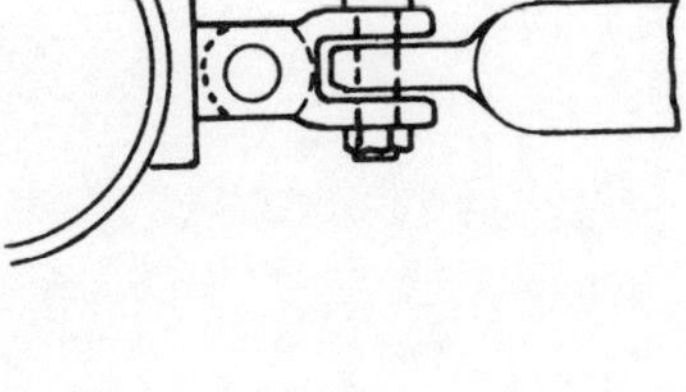

Local stresses are frequently found in the way of deck fittings and steps are taken to strengthen the plating in their vicinity,

MOORING BITTS are attached by a variety of methods, a very common method being either to bolt or weld directly to the deck.

PANAMA LEADS may be either fitted to the deck or in the bulwark as illustrated. It should be noted that the bulwark plating should be of increased thickness in way of the lead.

FAIRLEADS are of many types, a roller type being illustrated. They may be attached directly to the deck or to the deck and bulwarks.

'OLD MAN' OR 'DEAD MAN' is a type of fairlead used to prevent chafing and to give a direct lead of a mooring line to the windlass, winch or capstan.

MULTI ANGLE FAIRLEADS reduce the number of fairleads required for mooring, warping, locking and docking operations. They are typical of the type required for operations in the St. Lawrence Seaway.

EYEBOLTS AND RINGBOLTS and other attachments for the use of derrick gear, etc. are welded directly to the decks and bulwarks, the plating being stiffened accordingly.

WINCHES AND WINDLASSES; a lot of stress is caused in the vicinity of deck machinery and they are usually securely bolted to a base which is welded to the deck. Additional strength below decks including pillars is required in order that the stresses are spread over as wide an area as possible. Such mounting is shown on page 99.

DECK FITTINGS

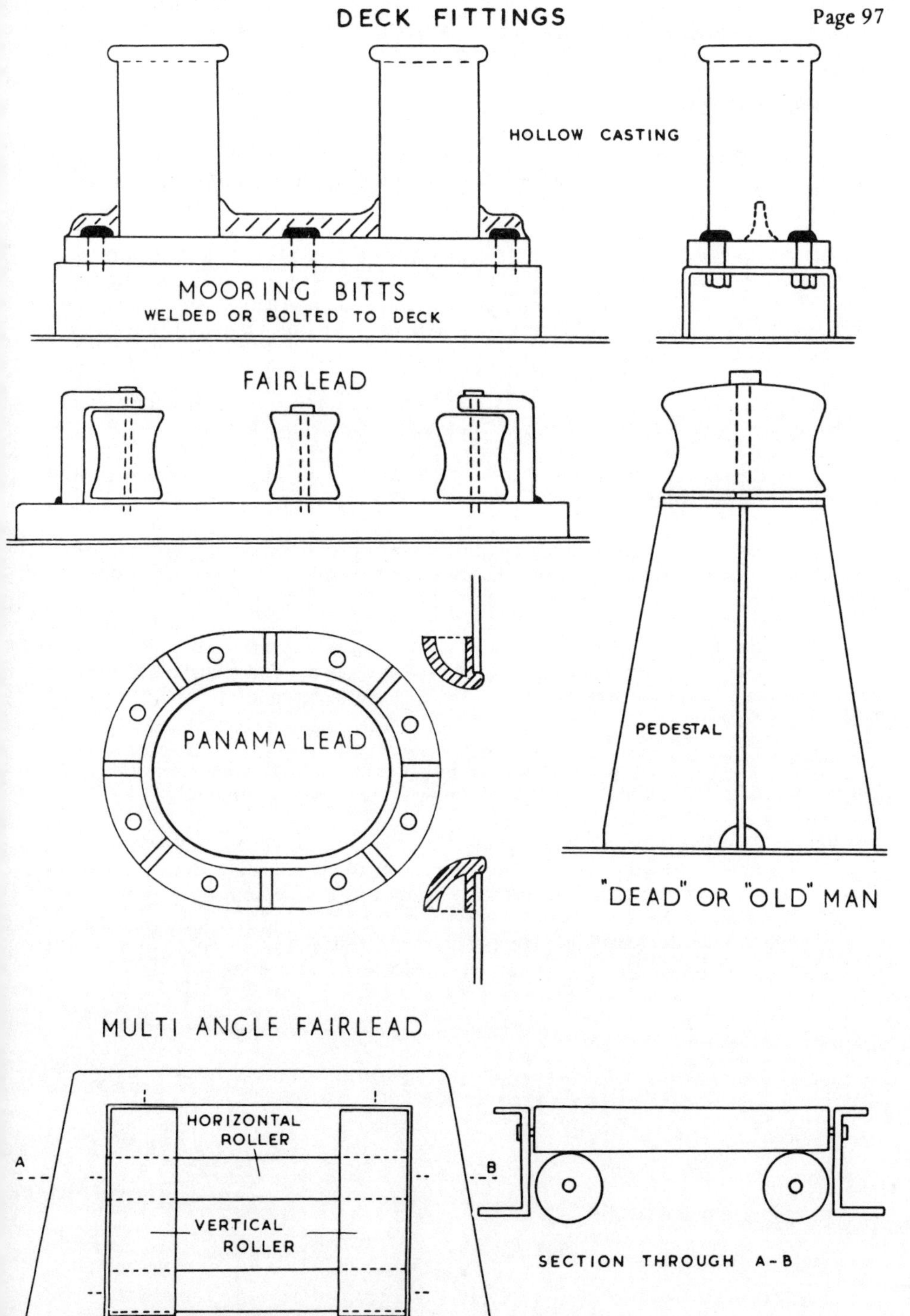
HOLLOW CASTING
MOORING BITTS
WELDED OR BOLTED TO DECK
FAIRLEAD
PANAMA LEAD
PEDESTAL
"DEAD" OR "OLD" MAN
MULTI ANGLE FAIRLEAD
HORIZONTAL ROLLER
VERTICAL ROLLER
A
B
SECTION THROUGH A-B

ENGINEROOMS, ETC.

The illustrations show the arrangement through an engineroom which is situated amidships and one that is situated aft. A section through the engineroom of a twin screw passenger vessel is shown on page 127.

The location of the engineroom is dependent on a number of factors such as the type of ship, number of screws, type of machinery etc. In cargo and passenger vessels it is usual to locate the machinery amidship or just abaft amidships. In bulk carriers i.e. tankers, ore carriers, etc. the machinery is located aft. The Rules require the engine room of a tanker to be situated aft.

The main engine seating is to be integral with the double bottom structure, the tank top plating in way of the seating being substantially increased in thickness.

Boiler bearers are to be of substantial construction, but in order to allow for expansion the connection of the boilers to the bearers is not rigid.

Adequate transverse stiffening is required throughout the double bottom, solid or plate floors at every frame, additional side girders to give the necessary support and strength.

Additional transverse strengthening is to be provided by means of web frames and strong beams with suitable pillaring or other arrangements. The webs are to be spaced not more than 5 frame spaces apart and are to have a depth of at least 2½ times the depth of the frame.

Where the machinery is aft the double bottoms are to be framed transversely. Webs as above are to be fitted whether the side framing is transverse or longitudinal.

In the machinery spaces two means of escape, one of which may be a watertight door, shall be provided from each engineroom, boiler room and shaft tunnel. There must be two means of communication between the bridge and engineroom.

Skylights have decreased in size and in many vessels are no longer fitted. Where fitted they should be adequately secured and protected by coamings.

ENGINEROOM

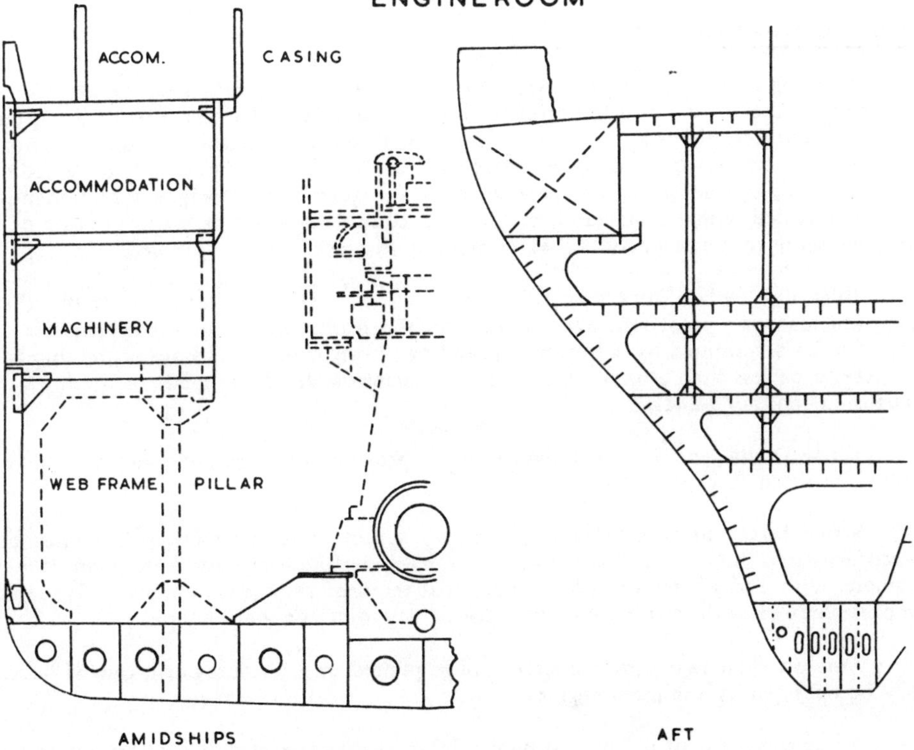

ACCOM.
CASING
ACCOMMODATION
MACHINERY
WEB FRAME
PILLAR
AMIDSHIPS
AFT

MACHINERY BOLTED TO BED
BRACKET
WINCH WINDLASS BED
WELDED TO DECK

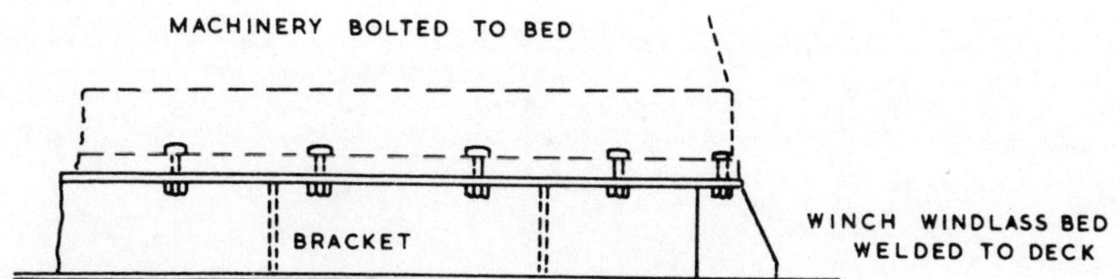

PUMPING AND PIPING ARRANGEMENTS

Pumping and piping arrangements in every ship should, in general, be capable of discharging water from any compartment when the ship is on an even keel or listed not more than 5 degrees either way. In machinery spaces, additional arrangements are required so that any water may be discharged through at least two bilge suctions, one connected to the bilge main and one to an independent pump or ejector. An emergency suction must also be provided with a connection to the main circulating water pump in the case of a steam vessel or to the main cooling water pump in the case of a motor vessel.

Bilge and ballast lines may be constructed of cast iron, steel, copper or other approved material. Heat sensitive materials such as lead must not be used. The size of bilge lines is determined by a formula depending on the main dimensions of the ship but is never to be less than 50mm. Provision for expansion should be made in the form of expansion bends or glands.

Screw Down Non-Return Valves must be provided on bilge lines, and mud boxes should be fitted to protect the pumps.

Strum boxes are not fitted in machinery spaces or tunnel sections but must be fitted in cargo holds. These strum boxes have perforations not more than 10mm diameter with a total area of at least twice that required for the suction pipe. It should be possible to clear the strums without breaking any joint in a suction pipe.

Not less than two power operated bilge pumps are to be provided, one of which may be operated by the main engines.

An arrangement of piping and pumps in an engineroom is shown diagrammatically opposite. The bilge system should only be designed to discharge overboard, but the ballast system is capable of discharging overboard, running up tanks by gravity and if necessary, pumping up tanks.

The fore peak tank has to have a screw down valve operated from above the bulkhead deck. This valve is to be inside the tank.

PUMPING AND PIPING

BILGE AND BALLAST LINES — DIAGRAM

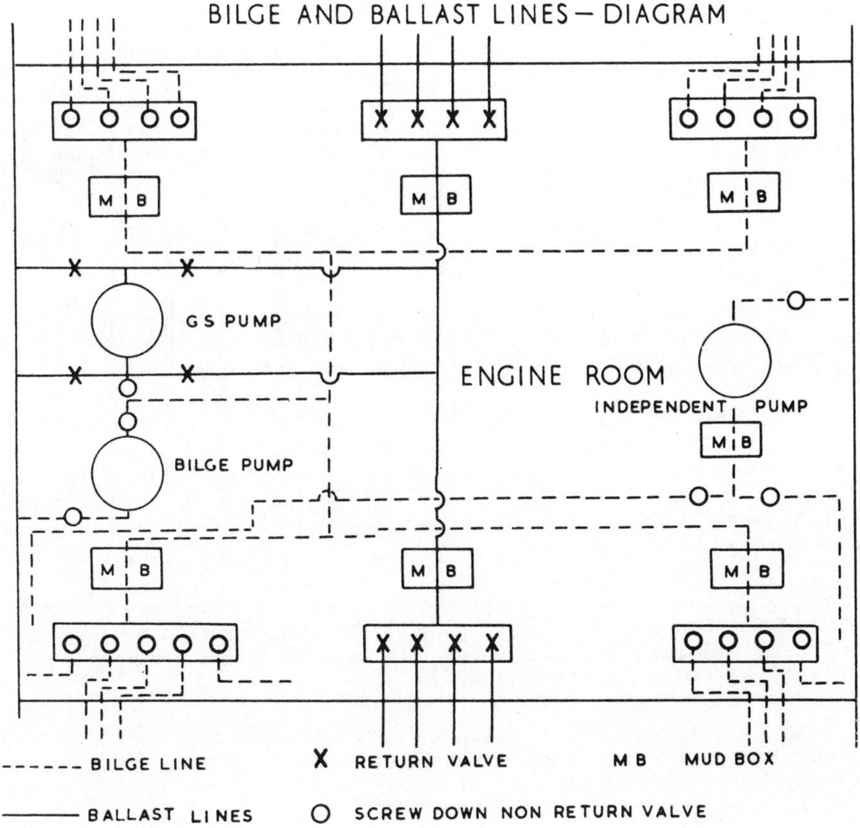

M B

M B

M B

G S PUMP

ENGINE ROOM

INDEPENDENT PUMP

M B

BILGE PUMP

M B

M B

M B

------ BILGE LINE

X RETURN VALVE

M B MUD BOX

——— BALLAST LINES

O SCREW DOWN NON RETURN VALVE

STRUM BOX

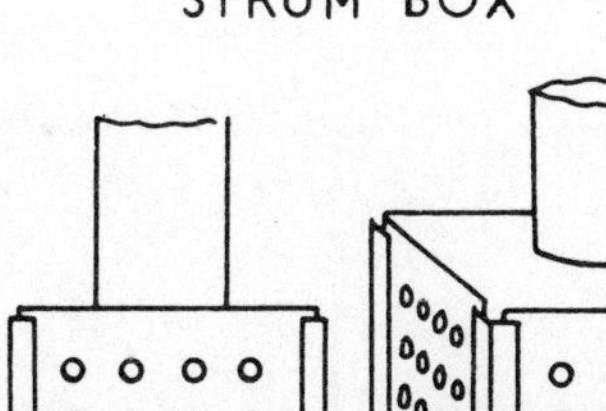

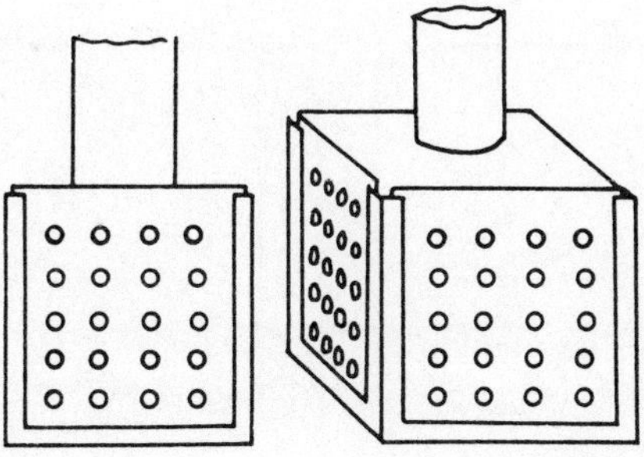

A GENERAL 'DRY' CARGO SHIP

The vessel illustrated has longitudinal framing at the decks and in the double bottoms, transverse framing at the sides. This arrangement is that recommended in the Rules, see page 42.

The direct welding of the frames to the shell plating may give rise to the rippling of the shell plating. Attention must be paid to the welding sequence and rippling may be reduced by scalloping the frames.

To provide the necessary degree of transverse strength transverses are fitted at the decks, see page 42, and plate floors fitted in the double bottoms, see page 38.

Longitudinal framing is not usual at the sides of general cargo vessels since this would necessitate the fitting of deep transverses 3.7 metres apart (to restore transverse strength) and would give rise to a large amount of broken stowage.

In cargo vessels ceiling is to be laid over the bilges and under hatchways; the ceiling over the bilges is to be arranged with portable sections that are easily removable. Where the tank top extends to the ship's side, as illustrated, bilge wells are fitted of not less than 0.17 cubic metre capacity. Where no ceiling is fitted the tank top plating is increased 2mm in thickness. Minimum thickness of tank top plating is laid down in the Rules where fork lift trucks are to be used.

Where a ceiling is laid in the square of the hatch it is to be not less than 65mm thick in the case of a wood ceiling and should be laid directly on the inner bottom plating, being embedded in a suitable composition, or laid on battens providing a clear space for drainage of at least 12.5mm

Cargo battens are to be fitted in the holds from above the upper part of the bilge to the under side of the beam knees and in all cargo spaces in 'tween decks and superstructures. Wood cargo battens are to be 50mm in thickness, the clear space between rows is not to exceed 230mm.

The vessel illustrated is fitted with partial centre line bulkheads in the lower holds and 'tween decks, extending from the transverse watertight bulkheads to the hatch coamings. This arrangement is very suitable for the carriage of grain and similar cargoes.

GENERAL CARGO SHIP

5TD
5LH
ER
4TD
4 LH
3
DT
2 3 TD
2 LH
1 TD
1 LH

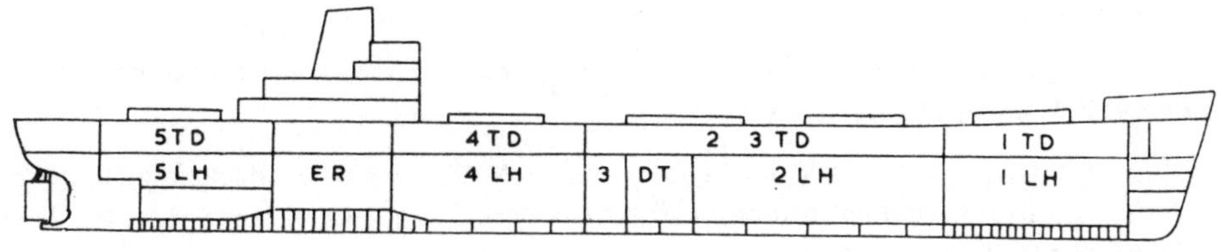

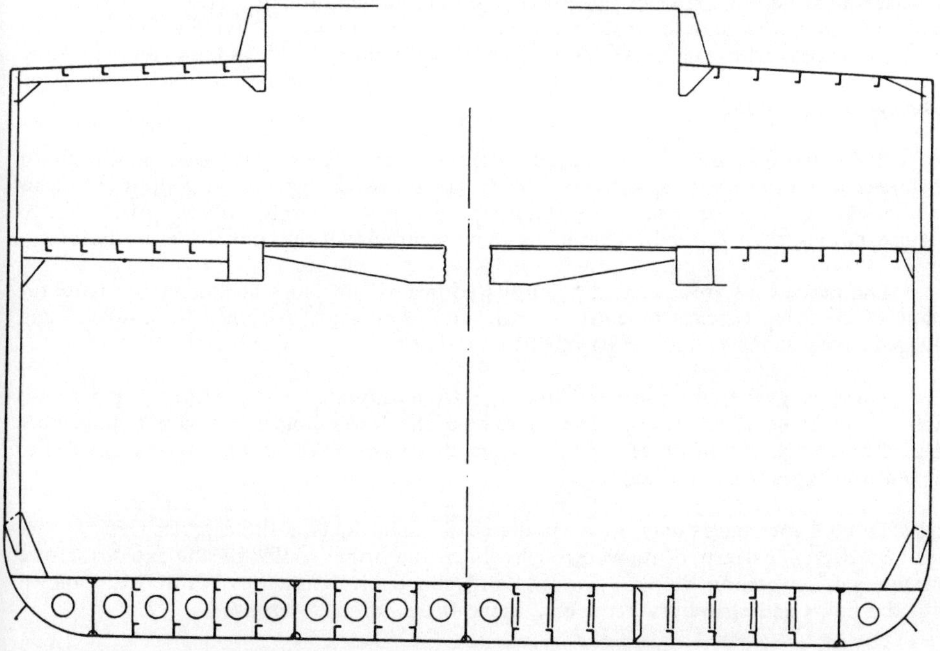

PLATE FLOOR
BRACKET FLOOR

REFRIGERATED VESSELS

These vessels are used for the carriage of cargoes which would deteriorate at ordinary hold temperatures.

To facilitate the fixing of insulation, transverse framing at the decks may be fitted in preference to longitudinal framing and transverses. Longitudinal framing is still fitted in the double bottom.

The holds and 'tween decks are insulated by packing an insulating material (fibre-glass, silicate of cotton, slab or granulated cork) between the frames. This is held in place by wood sheathing or galvanised sheeting, as illustrated. The latter arrangement is that most frequently met in modern refrigerated vessels. The deckhead is insulated in a similar manner. Tank top insulation (illustration on page 41), is slightly different in that a 50mm air space must be left between the crown of an oil tank and the insulation, also 50mm of hardwood sheathing should protect the insulation in the way of hatchways. Insulated plugs are placed on bilge limbers. Alternative arrangement to an air gap is a layer, 12.5mm thick, of an inodourous material supplied to the tank top or a 5mm thick layer of oil resisting composition.

Cargo battens are to be secured to the insulating linings at sides, bulkheads of chambers and other vertical surfaces. They are to be arranged to suit the flow of air and should generally be 50mm by 50mm spaced approximately 400mm apart. Cargo battens, not less than 75mm by 75mm, are to be secured over the tunnel top insulation.

The hatchways are insulated by insulated beams and plugs as illustrated. Note the shape of the plug hatches to ensure a tight fit. Any ventilator must be insulated and plugged, likewise masts and pillars must be insulated.

There is to be provision for drainage from insulated compartments and this is effected by brine sealed traps. The pipe from the lower hold must have a non-return trap. These traps are to prevent odours from the bilges reaching the cargo spaces and to prevent the bilges from freezing.

To cool the spaces cold air is circulated through ducts in the cargo spaces. The air is drawn over a 'battery' of pipes through which cold brine (cooled by the expansion of a suitable gas i.e. carbon dioxide, Freon 12, Argon 6 etc) circulates before being blown into the holds and chambers, fans, etc., being fitted for this purpose.

Surveys are held before each cargo is loaded, when the surveyor checks the cargo spaces for cleanliness and sound insulation and the temperatures are noted. The refrigerating machinery is examined under working conditions.

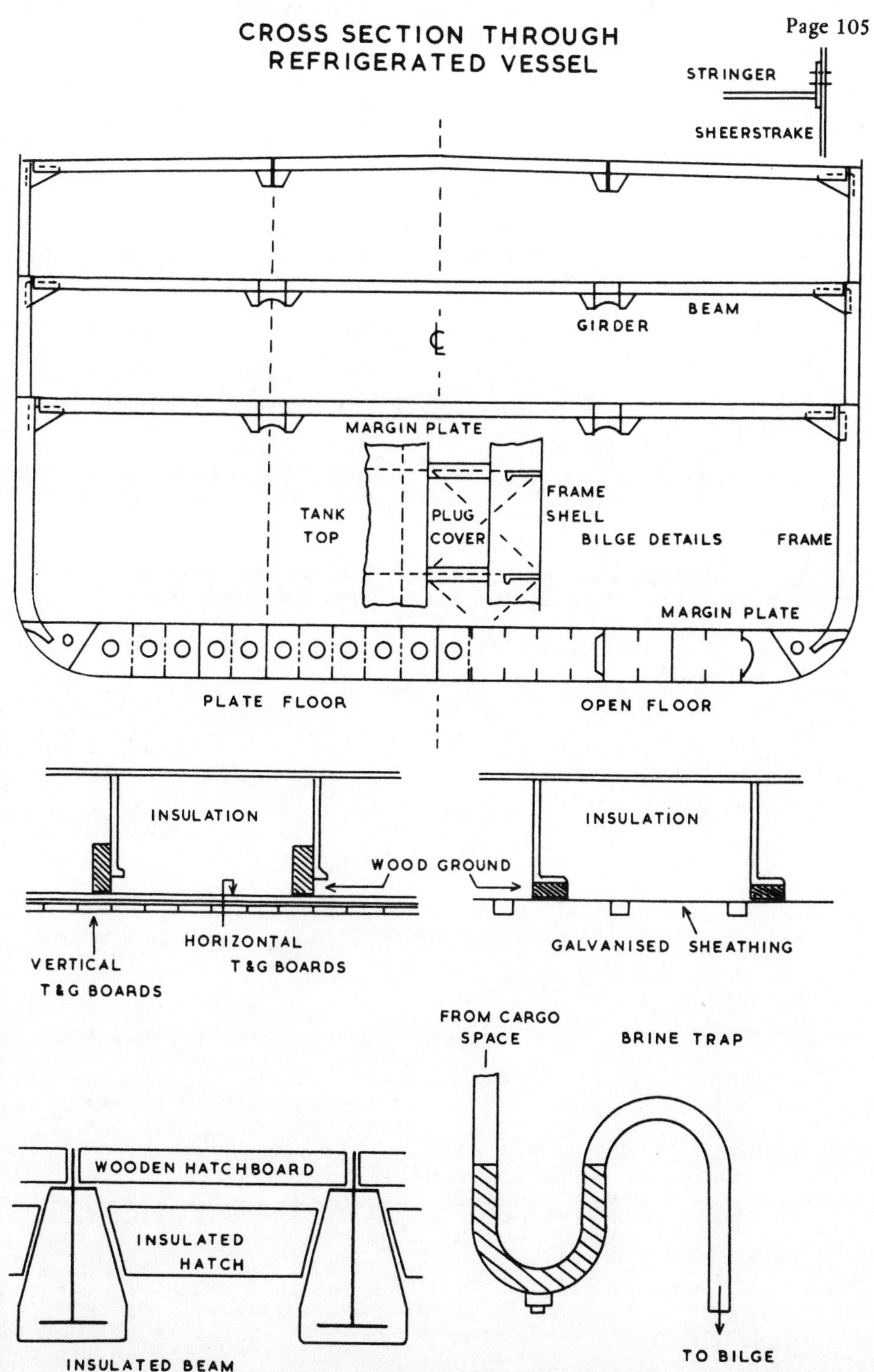
CROSS SECTION THROUGH
REFRIGERATED VESSEL
STRINGER
SHEERSTRAKE
BEAM
GIRDER
℄
MARGIN PLATE
FRAME
SHELL
TANK
TOP
PLUG
COVER
BILGE DETAILS
FRAME
MARGIN PLATE
PLATE FLOOR
OPEN FLOOR
INSULATION
WOOD GROUND
HORIZONTAL
T&G BOARDS
VERTICAL
T&G BOARDS
INSULATION
GALVANISED SHEATHING
FROM CARGO
SPACE
BRINE TRAP
WOODEN HATCHBOARD
INSULATED
HATCH
INSULATED BEAM
TO BILGE

DRY CARGO/CONTAINER CARRIER

The vessel illustrated is capable of loading break bulk cargo, pallets, and containers. The upper deck is suitably stiffened, and hatches constructed accordingly, for the carriage of containers on deck.

Side doors are fitted at the upper 'tween deck level to facilitate loading direct from the quay to the ship.

Sliding bulkhead doors are fitted connecting the holds in the 'tween decks and upper 'tween decks. They are pneumatically operated.

Flush fitting hatch covers are fitted in the lower decks to permit the use of fork lift trucks.

The height of the vessel has been carefully designed so as to permit the stowage of five tiers of containers.

Large centre line pillars are fitted in conjunction with a centre line girder and cantilevers at the ship's sides. The spacing of pillars and cantilevers is as illustrated in the profile.

The holds are sealed by watertight electrically driven hatch covers on the weather decks, stowage for the covers being arranged at each end of the hatch.

Vessels of this type are frequently fitted with electrically driven deck cranes to serve the various holds.

DRY CARGO - CONTAINER CARRIER

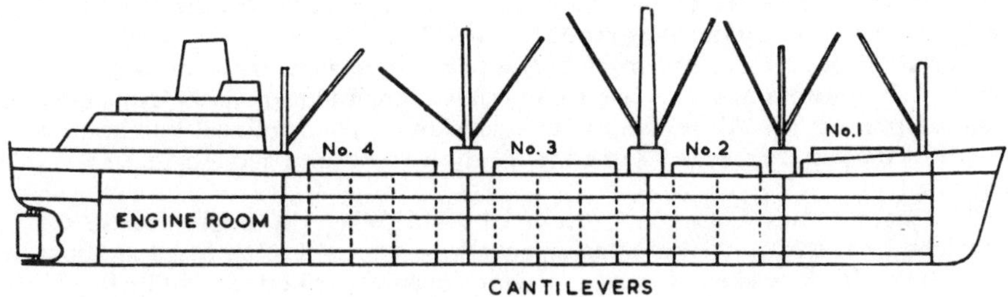

No. 4
No. 3
No. 2
No. 1
ENGINE ROOM

CANTILEVERS

ENGINE CASING
No. 4 HATCH
No. 3 HATCH
No. 2 HATCH
No. 1 HATCH

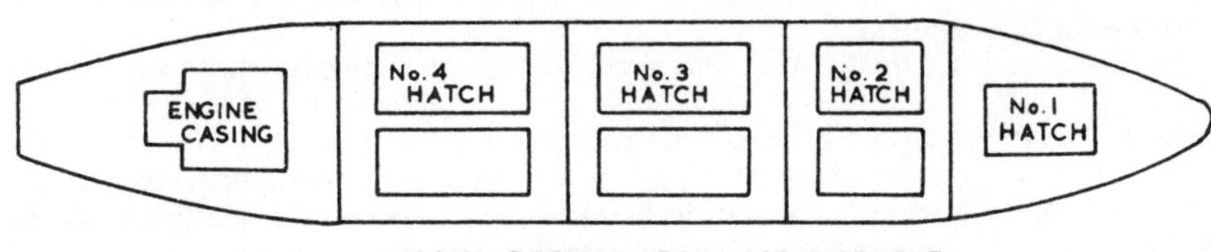

MAIN DECK HATCH ARRANGEMENT

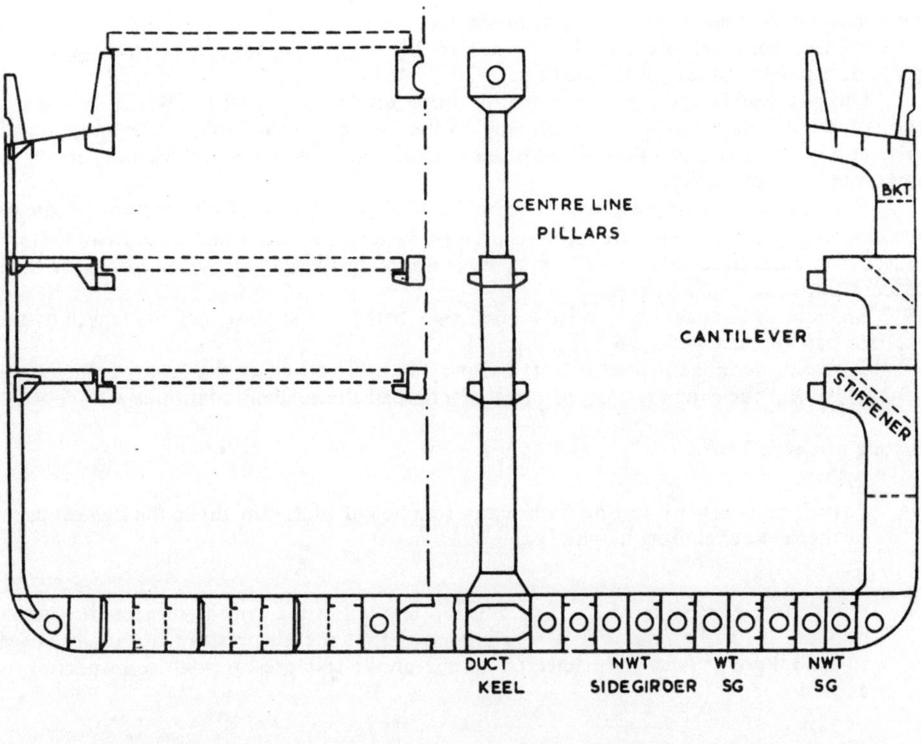

CENTRE LINE PILLARS
BKT
CANTILEVER
STIFFENER
DUCT KEEL
NWT SIDEGIRDER
WT SG
NWT SG

TANKERS

Readers will be aware of the many special features of these vessels and reference should be made to the various illustrations.

They are single deck vessels, machinery aft and fitted with two or more longitudinal bulkheads. The bottom and deck are to be framed longitudinally in the cargo tanks and longitudinal framing should be used at the sides and on longitudinal bulkheads where the length of the vessel exceeds 200 metres. Both the combined and longitudinal systems are illustrated.

The length of any tank is not to exceed 0.2L. When the length exceeds 0.1L or 15m, whichever is the greater, a transverse wash bulkhead is to be fitted at about mid length of the tank. A centre line bulkhead, which may be perforated, is to be fitted when the breadth of the centre tank exceeds 18.3m (breadth of vessel less than 35.05m) and 11.29 + 0.2B metres (breadth of vessel greater than 35.05m).

Cofferdams are to be provided at the forward and after ends of the oil cargo spaces; they are to be at least 760mm in length. Pumprooms and water ballast tanks will be accepted in lieu of cofferdams.

The Rules relating to the deck and shell plating are similar to those for dry cargo vessels.

The end connections of bottom, side and deck longitudinals to bulkheads are to be of such a nature as to ensure adequate fixity and continuity of longitudinal strength.

The spacing of bottom, side and deck transverses in general is not to exceed 3.6m and the depth of a transverse should not be less than 2.5 times the depth of the slot cut for the longitudinal.

Where transverse side framing is adopted, side stringers are to be fitted as required, the number is dependent on the depth of the tank.

Where cross ties are fitted they are to be connected to vertical transverses or horizontal girders by suitable brackets.

Oiltight bulkheads may be plane or horizontally corrugated. Where the ship's side is framed longitudinally the stiffening on the longitudinal bulkhead is to be arranged horizontally. Transverse oiltight bulkheads may be plane or with corrugations arranged horizontally or vertically.

Corners of oiltight hatchways are to be well rounded with a minimum height of coaming of 250mm. Covers are to be secured by fastenings spaced not more than 457mm apart in a circular hatchway or 380mm apart in a rectangular hatchway and not more than 230mm from the hatch corners.

Ships with bulwarks are to have open rails fitted for at least half the length of the exposed part of each well.

A strong permanent fore and aft gangway is to be fitted at the level of the superstructure deck; the gangway may be omitted when all the accommodation is aft.

Testing of Cargo Tanks.

Either

(1) A structural test by testing with water to a height of 2.45m above the highest point of the tank, excluding hatchways.

or

(2) A leak test consisting of a soapy solution test while the tank is subjected to an air pressure of $0.14 kg/cm^2$. It is recommended that the air pressure is initially raised to $0.21 kg/cm^2$ and then lowered to the above test pressure before inspection is carried out.

TANKER - COMBINED SYSTEM

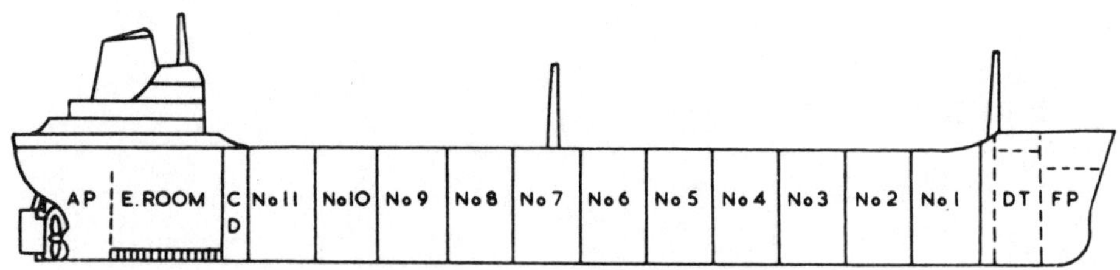

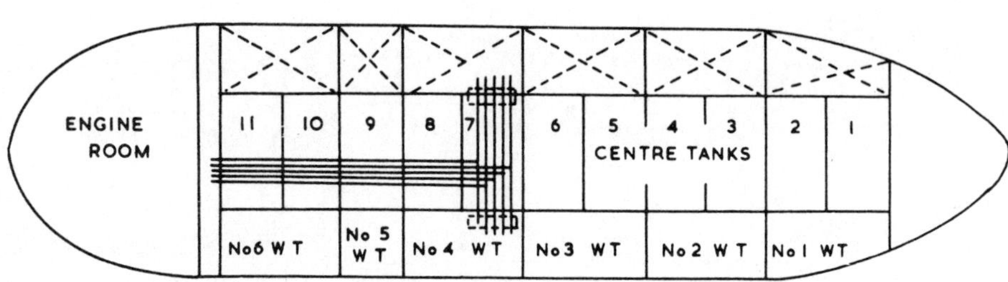

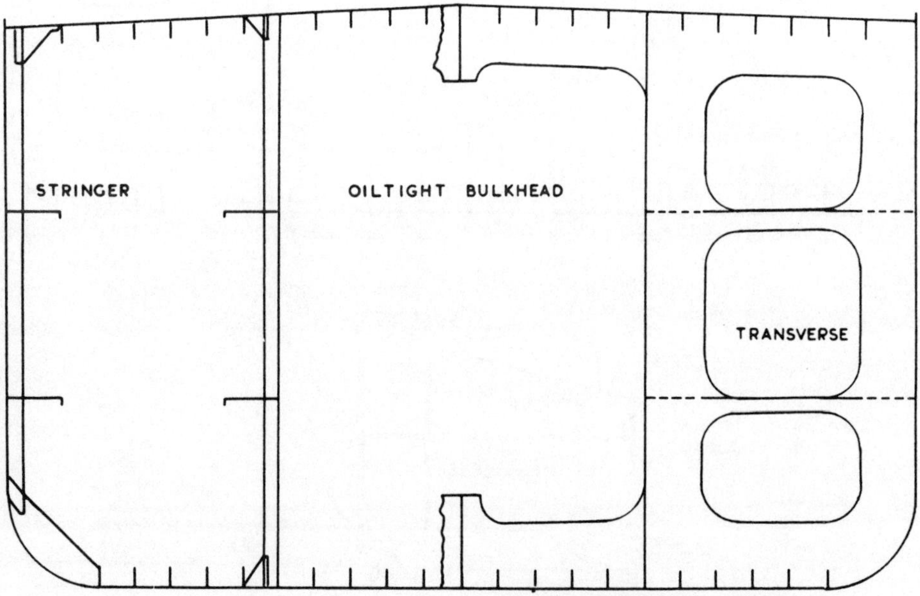

STRINGER

OILTIGHT BULKHEAD

TRANSVERSE

ARRANGEMENT EVERY 4TH FRAME

TANKER- LONGITUDINAL SYSTEM

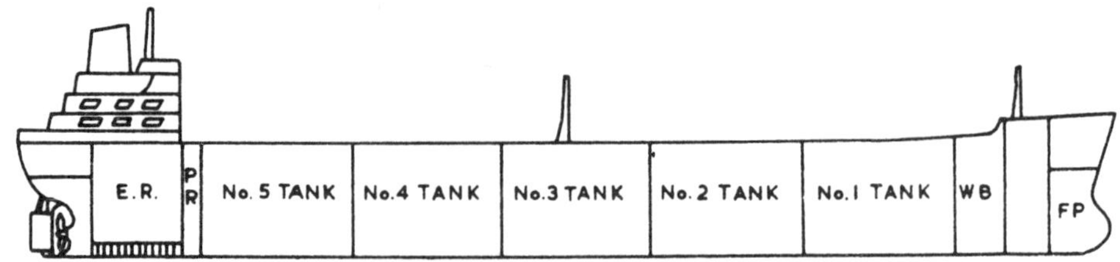

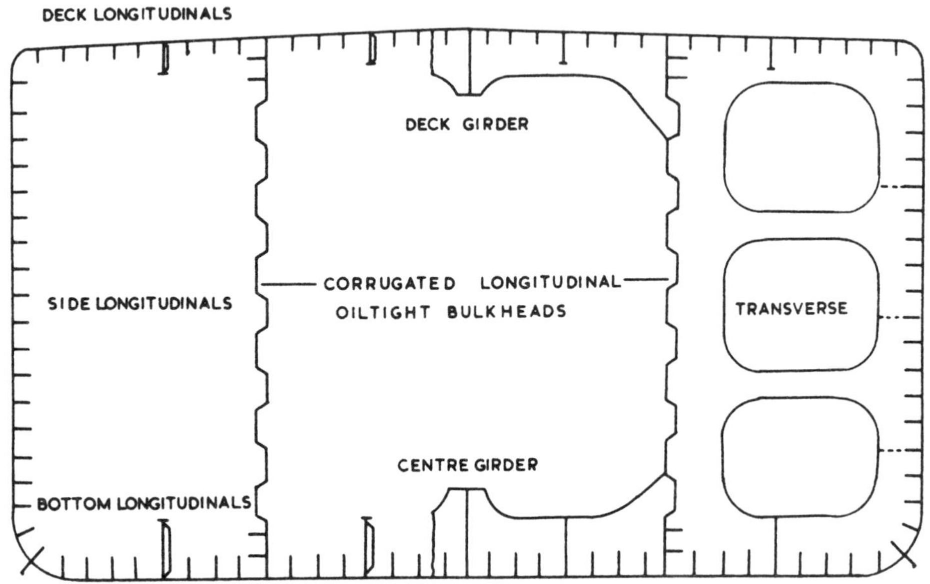

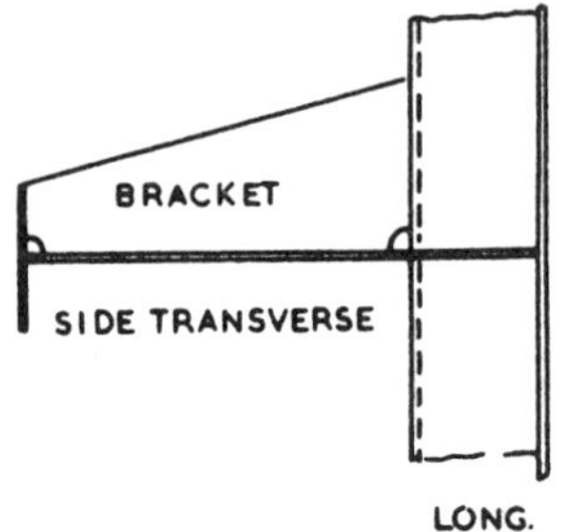

DETAIL OF TRIPPING BRACKET
ON SIDE TRANSVERSE

V. L.C.C. TANKER

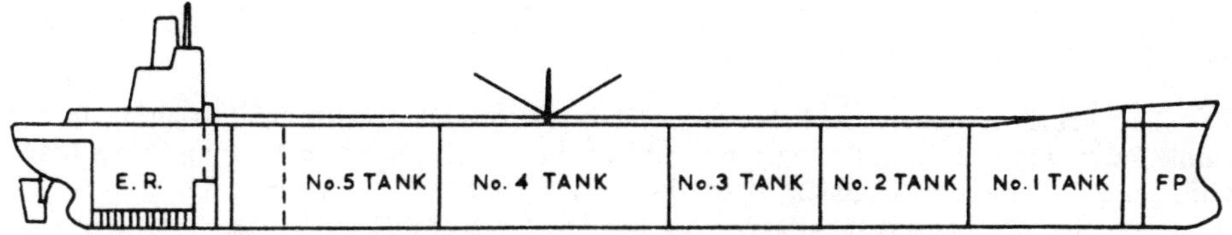
E. R.
No.5 TANK
No. 4 TANK
No.3 TANK
No. 2 TANK
No. I TANK
FP

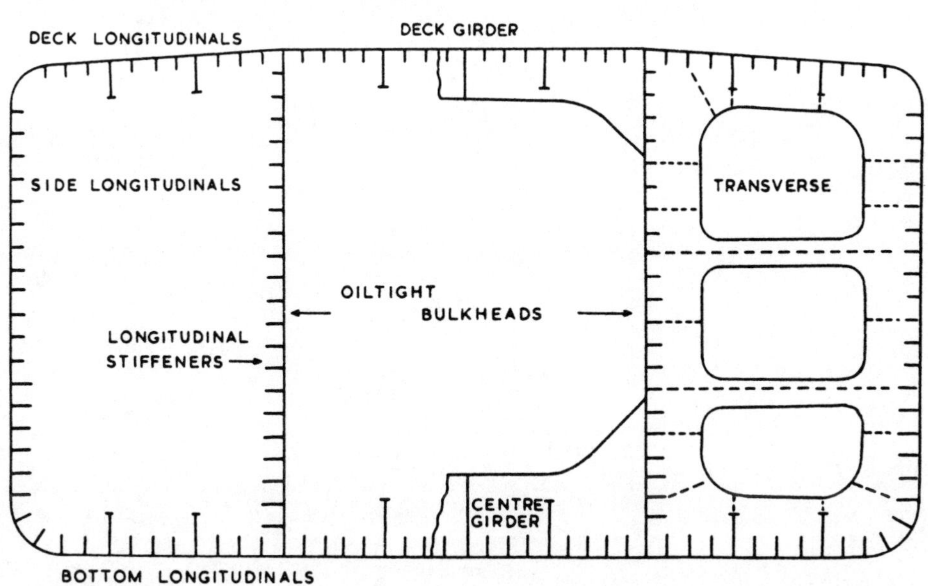
DECK LONGITUDINALS
DECK GIRDER
SIDE LONGITUDINALS
TRANSVERSE
OILTIGHT
BULKHEADS
LONGITUDINAL
STIFFENERS
CENTRE
GIRDER
BOTTOM LONGITUDINALS

BULK CARRIER

This type of vessel is designed to load a maximum deadweight of any type of bulk cargo, from heavy ore to light grain. The vessel illustrated has been designed to carry bulk sugar as the main commodity.

The vessel is constructed on the combined system, longitudinal framing in the double bottoms, bottom of wing tanks and at the deck, transverse framing being fitted at the sides. Transverse webs are fitted in the wing tanks at intervals of 3.4 metres, side stringers being fitted at approximately one third and two thirds the depth of the tanks.

The vessel has two longitudinal watertight bulkheads thus giving the minimum freeboard for this type of vessel (i.e. B-100).

The wing tanks may be used for the carriage of grain, bulk cargo or water ballast.

BULK CARRIER

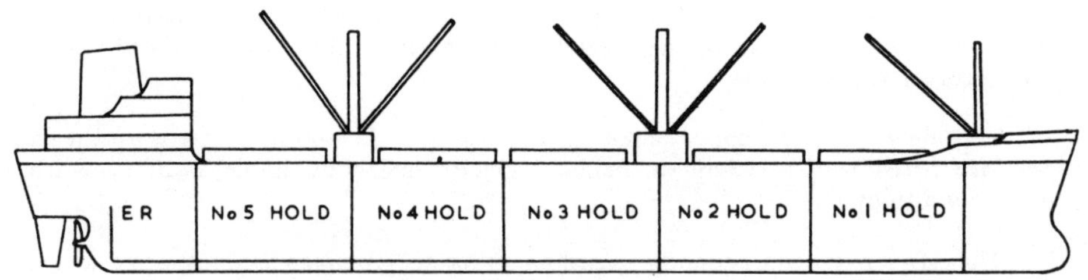

GRAIN, BULK CARGO, WATER BALLAST CARRIED IN WING TANKS

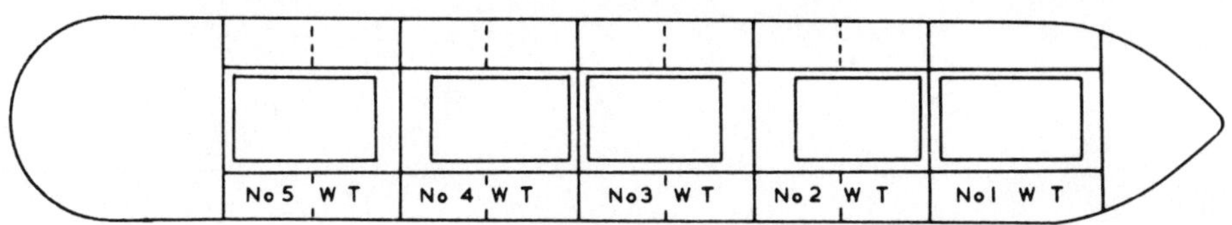

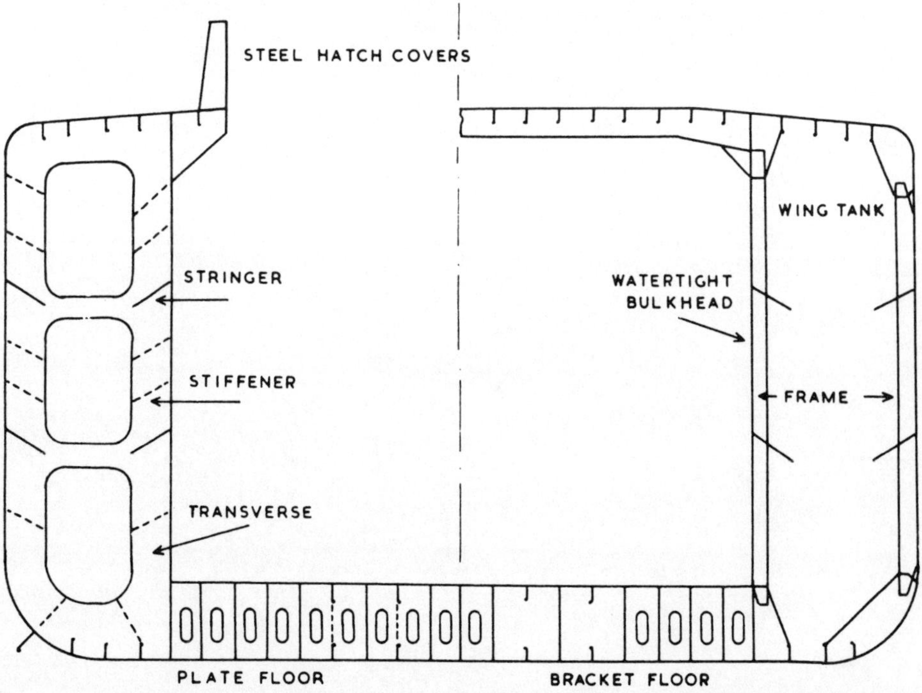

STEEL HATCH COVERS

STRINGER

STIFFENER

TRANSVERSE

WING TANK

WATERTIGHT BULKHEAD

FRAME

PLATE FLOOR

BRACKET FLOOR

A MODERN COLLIER OR BULK CARRIER

This type of vessel is designed for the rapid loading and discharging of cargo. To achieve this very wide, long and high hatchways are fitted. Hatch covers are of steel with direct securing arrangements.

Topside tanks and hopper tanks (an extension of the double bottoms up the ship's sides) are fitted to give adequate ballast capacity and thus ample stability when in ballast conditions.

The after deck is frequently raised in order to give increased capacity and thus prevent any tendency to trim by the head when loaded.

The mast may be telescopic in order that low bridges may be safely negotiated.

A profile and cross section of a modern collier is illustrated opposite.

COLLIER

AP
ENGINE
No. 4 HOLD
No. 3 HOLD
No. 2 HOLD
No. 1 HOLD
FP

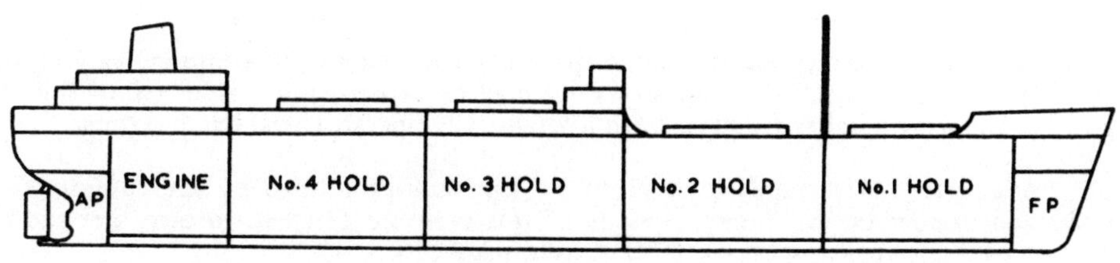

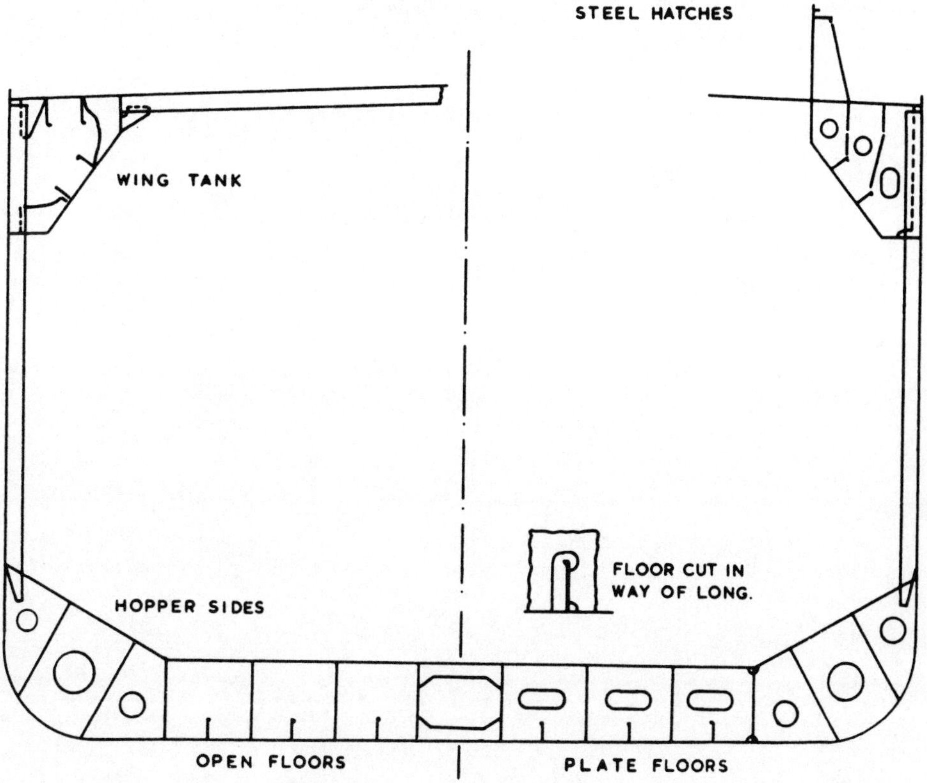

STEEL HATCHES
WING TANK
FLOOR CUT IN WAY OF LONG.
HOPPER SIDES
OPEN FLOORS
PLATE FLOORS

BULK ORE CARRIERS

These are designed to carry a high density cargo, and the particular requirements of this trade are for the vessel which has adequate strength for the heavy loading, a centre of gravity high enough to prevent undue stiffness when fully loaded and arrangements to ensure maximum speed in loading and discharging.

The construction of bulk carriers varies considerably, the type shown being that generally built.

The type illustrated has its cargo space divided into six self trimming cargo holds and is so designed that full cargoes of grain may be carried without the use of shifting boards. The wide hatches are fitted with hydraulically operated steel hatch covers.

Adequate ballast capacity is given by the tankage formed by the topside wing tanks and hopper tanks. Some versions of this type are designed to carry ore and oil cargoes on alternate voyages, see OBO carrier, page 119.

The type shown has the following particulars, L = 160m, B = 23m, D = 13m, draught = 9.4m, Deadweight = 22000 tonnes. About 10000 tonnes of water ballast may be carried, No. 3 hold being constructed as a deep tank.

ORE CARRIER

6 Hold
5 Hold
4 Hold
3 Hold
2 Hold
1 Hold

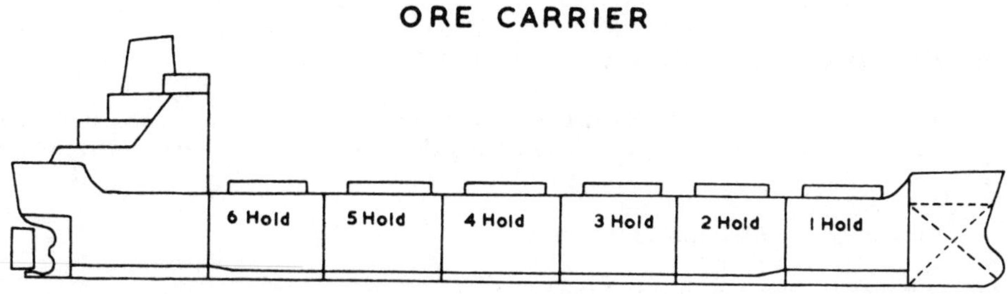

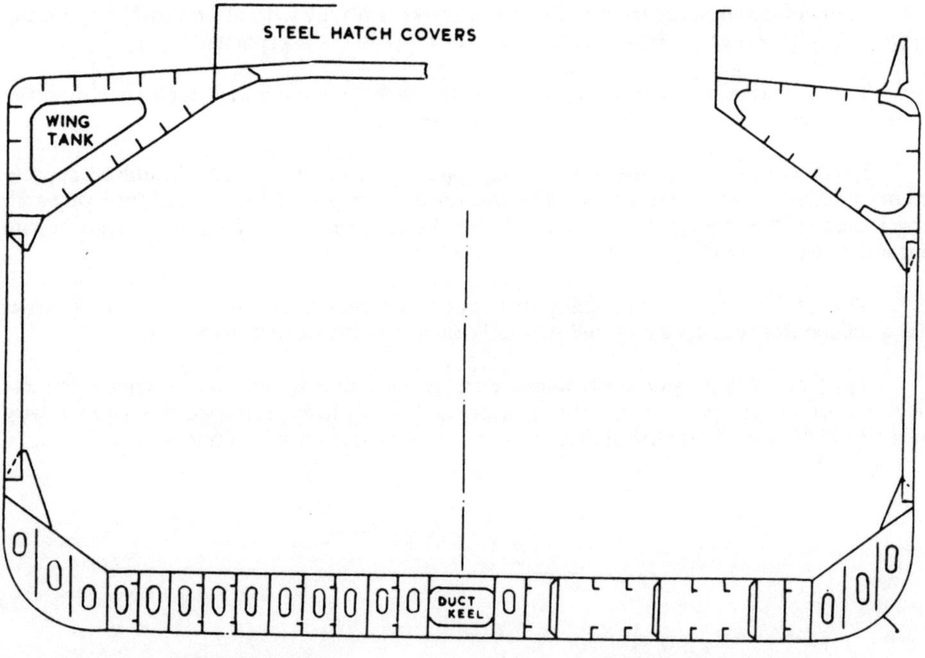

STEEL HATCH COVERS
WING TANK
DUCT KEEL
PLATE FLOOR
BRACKET FLOOR

OBO (ORE/BULK/OIL) CARRIER

The major design characteristics peculiar to this type of vessel as compared to an ore carrier is the double skin at the sides having all the stiffening within the narrow wing tanks.

Advantages of this double skin are :—

(1) it makes for easier cleaning of the holds,

(2) the inner skin reduces free surface in the large cargo holds,

(3) the clean ballast capacity of the vessel is increased.

Transverse bulkheads are usually of the cofferdam type with all the stiffening on the inside. Though easier to clean there is however a loss in cubic capacity.

There is usually a rise in the floor of the inner bottom which facilitates drainage to drain recesses or wells arranged on the centre line.

Hatch covers are of the side rolling type. The hatch breadth should be approximately 50 per cent of the beam. The illustrations on page 79 show the type of hatch cover used in this type of vessel, gas tightness being essential. Hydraulic operation with automatic battening down is a feature of these hatch covers.

The OBO Carrier is technically different from the ore/oil carrier. The OBO carrier has a smaller deadweight compared to a bulk carrier of the same dimensions.

The flexibility of this type of carrier is of particular appeal to an owner who has a network of contracts covering the transportation of many commodities over a large area. An OBO carrier has the ability to switch readily from dry to liquid cargoes.

OBO CARRIER

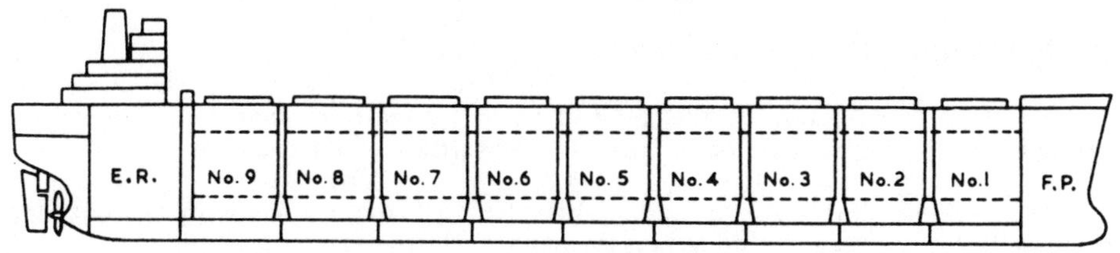

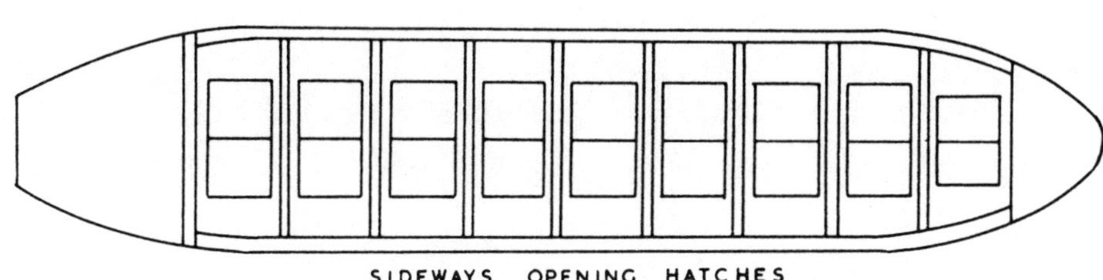

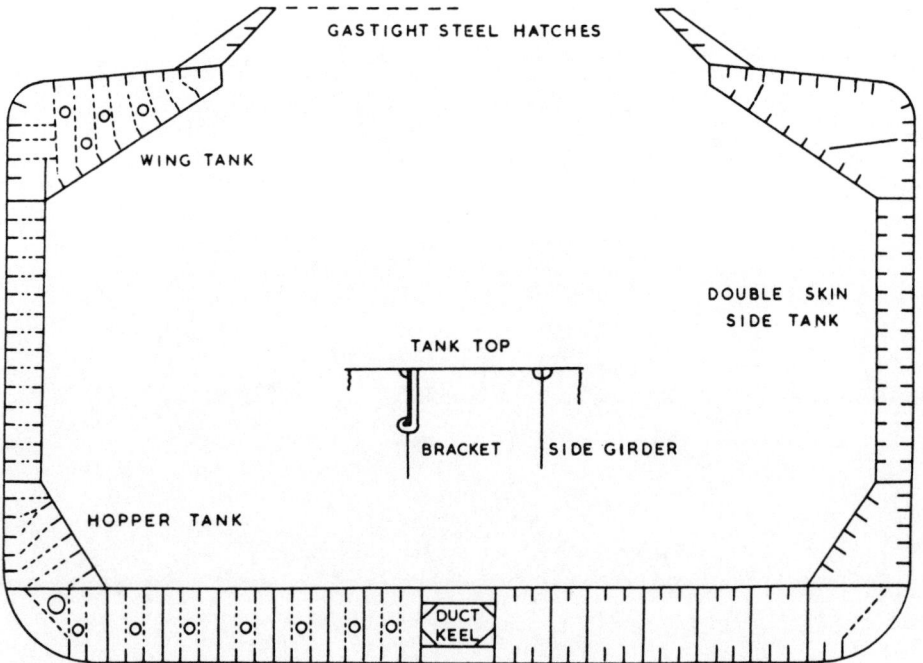

GASTIGHT STEEL HATCHES

WING TANK

DOUBLE SKIN SIDE TANK

TANK TOP

BRACKET

SIDE GIRDER

HOPPER TANK

DUCT KEEL

PLATE FLOOR

BRACKET FLOOR

LIQUID PETROLEUM GAS CARRIER

The hull construction of this type of vessel is similar to that of an ore carrier, topside wing and hopper tanks being fitted for the carriage of water ballast.

Longitudinal or transverse framing may be used at the sides with an inner hull occasionally being fitted. Both arrangements are illustrated.

The cargo of liquefied gases is carried in independent tanks. Free standing tanks are usually rectangular or trapezoidal in section with a hatch extending through the deck giving access to the pipes, pumps and gauges.

LIQUID PETROLEUM GAS CARRIER

PRESSURE VESSEL
ENGINE ROOM
No.4 GAS TANK
No. 3 GAS TANK
No. 2 GAS TANK
No. 1 GAS TANK

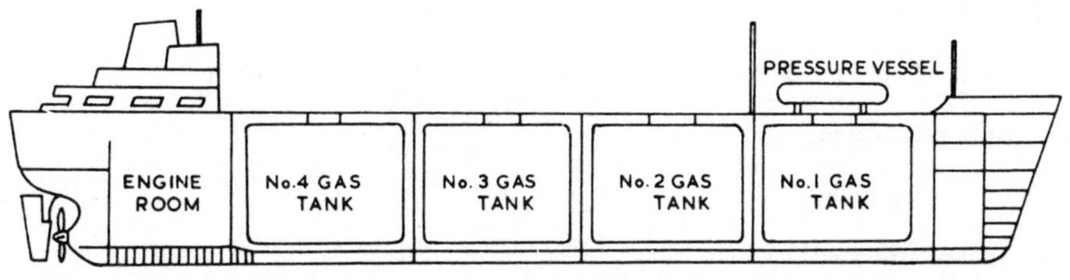

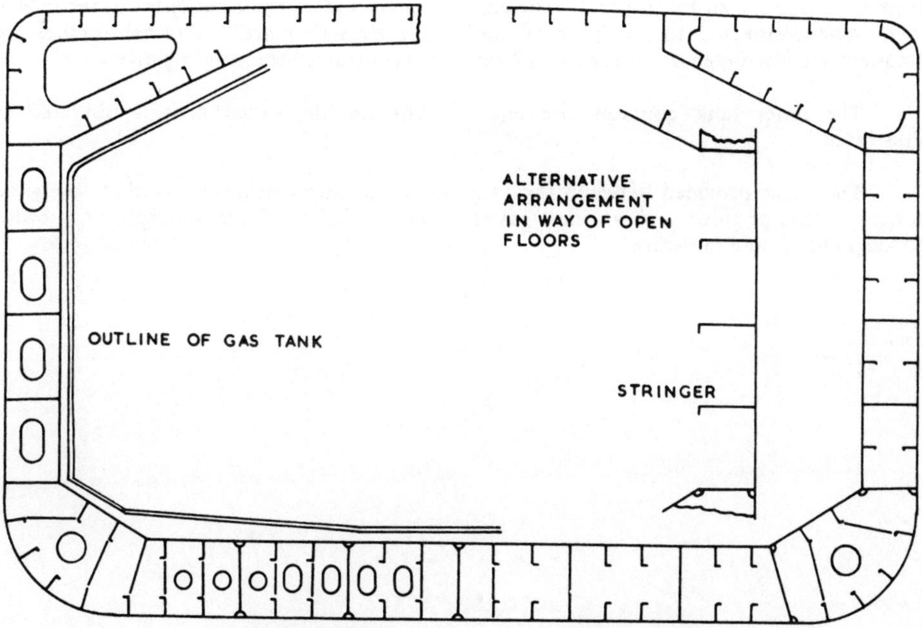

ALTERNATIVE ARRANGEMENT IN WAY OF OPEN FLOORS
OUTLINE OF GAS TANK
STRINGER
PLATE FLOOR
OPEN FLOOR

The cargo tanks are located by supports on the double bottom. Additional provision must be made to ensure that the cargo tanks will not move when the vessel is pitching and rolling in a seaway, also the gas tanks must not be permitted to float should the compartment (hold) become flooded when the tanks are empty. When cargo is carried at low temperatures provision is to be made for expansion and contraction of the tanks.

Liquefied gases may be transported in one of four ways, as follows :—

(1) fully pressurised,

(2) fully refrigerated,

(3) semi-refrigerated.

(4) fully pressurised/fully refrigerated.

The way in which the cargo is transported controls the construction of the tanks and the quality of the steel to be used in them.

The tanks are usually double walled consisting of a structural inner or primary membrane supported by internal webs and stiffeners and a structural outer or secondary membrane connected to the primary membrane by web plates. The membranes are frequently constructed of corrugated plating thus providing structural rigidity.

The inner tank contains the liquid. The double walled tank is insulated as illustrated.

The space provided between the cargo tank and ship's structure is either accessible for inspection or filled with an inert gas at a pressure slightly above atmospheric in order to keep out air and moisture.

LIQUID GAS CARRIER

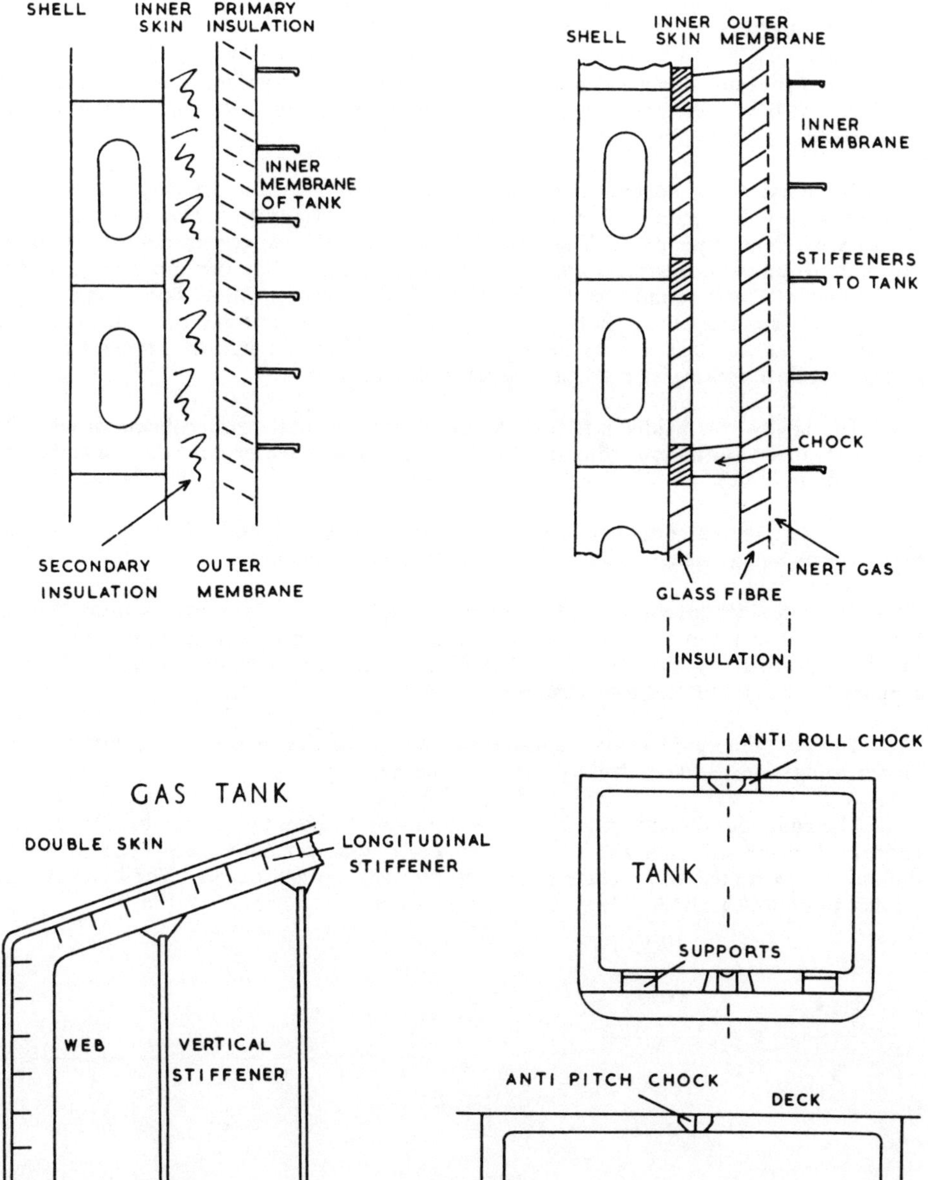
SHELL
INNER SKIN
PRIMARY INSULATION
INNER MEMBRANE OF TANK
SECONDARY INSULATION
OUTER MEMBRANE
SHELL
INNER SKIN
OUTER MEMBRANE
INNER MEMBRANE
STIFFENERS TO TANK
CHOCK
INERT GAS
GLASS FIBRE
INSULATION
GAS TANK
DOUBLE SKIN
LONGITUDINAL STIFFENER
WEB
VERTICAL STIFFENER
ANTI ROLL CHOCK
TANK
SUPPORTS
ANTI PITCH CHOCK
DECK
TANK
COLLISION CHOCK
SUPPORTS

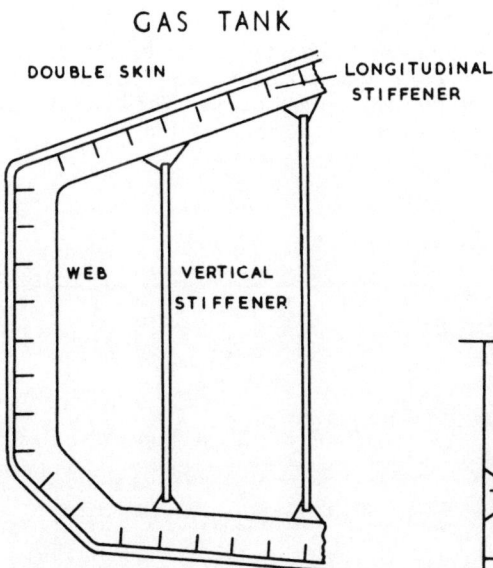

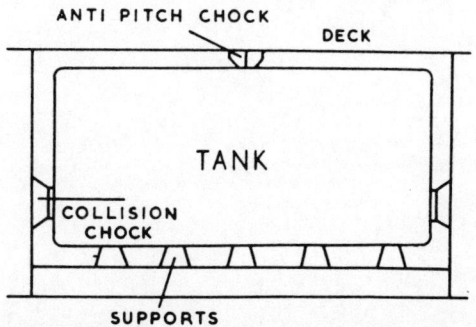

CONTAINERSHIPS

The main object in the design of these vessels is to carry the maximum number of containers within the designed length and breadth having regard to the form and structural arrangement.

The provision of adequate structural strength is of prime importance. Longitudinal framing is used throughout the main body of the vessel, transverse framing being used in the fore part and after part. They are built having a cellular construction at the sides. Strong longitudinal box girders are formed port and starboard by the upper deck, second deck, top of shell plating and top of the longitudinal bulkhead, i.e. inner skin or shell. High tensile steels are frequently used in the upper deck and sheerstrake to form the strong box girder. These box girders in addition to providing longitudinal strength provide stiffness against racking stresses as well as useful tank spaces.

The hatchway is divided into three sections, two long hatch girders being fitted. The girders are made continuous thus sharing the longitudinal bending strength and adding to the section modulus.

The carriage of containers on deck results in a high deck loading, the decks and hatch covers being strengthened to withstand this extra loading.

The container spaces are suited either for 12.20m or 6.10m units. A form of bulkhead is fitted at intervals of 14.70m, centre to centre with watertight bulkheads being fitted as required by the Rules. In addition to resisting racking the bulkhead give support to the double bottom structure.

The container guides and associated structures are designed to withstand dynamic (accelerating) forces due to rolling, pitching and heaving.

The guide consists of angle bars approximately 150mm by 150mm by 14mm thick connected to vertical webs and adjoining structure, spaced 2.60m apart. The bottom of the guides are bolted to brackets welded to the tank tip and beams. The brackets are welded to doubling plates, 15mm thick which are welded to the tank top.

CONTAINER SHIP

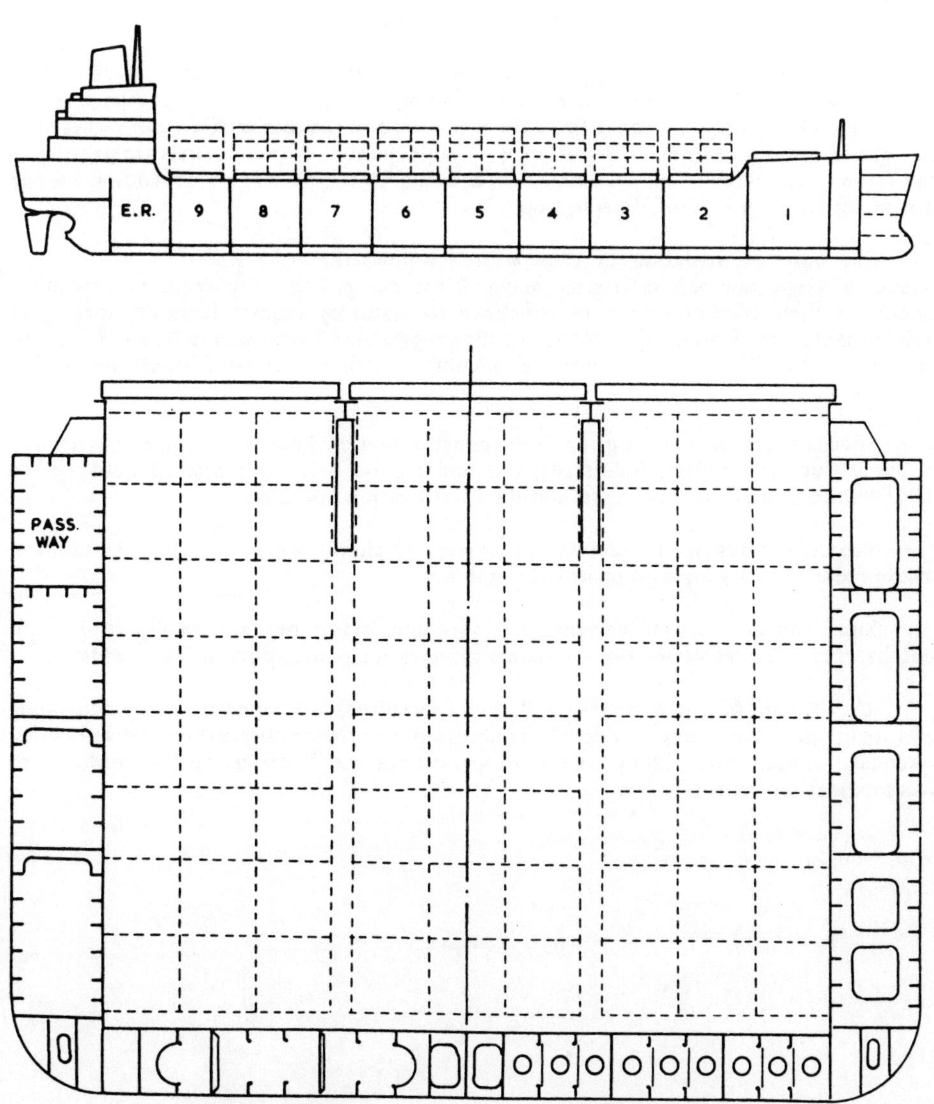

PASSENGER VESSELS

The illustrations show the profile and cross sections of a twin screw passenger vessel. The cross sections have been drawn in by way of a midship section, the engineroom and an after compartment. The engineroom is situated abaft amidships.

Any vessel which carries over 12 passengers is considered a Passenger Vessel under the provisions of the SOLAS Convention. Passenger vessels range from the large ocean liners and cruise ships with space for little or no cargo to the intermediate type where the carriage of cargo is of primary importance. A large number of these vessels are passenger/car ferries, the amount of accommodation available for passengers depending on the season, see Roll on/Roll off Vessels. page 129.

The basic construction of these vessels follows the same Rules for a dry cargo vessel, a large number of decks being fitted to provide the required amenities. Details of their construction with reference to statutory requirements i.e. spacing of bulkheads etc., are contained in Merchant (Passenger) Ship Construction Rules, Statutory Instrument No. 1013. They must also comply with the latest SOLAS and IMCO requirements.

Special attention must be given to the construction of these vessels, especially in way of superstructures, pillars, bulkheads, etc. and a close liaison maintained between the naval architects designing the vessel and the Classification Society.

Superstructures are frequently constructed of aluminium alloys which in addition to the reduction of weight improves the stability.

Since the comfort of passengers is very important they are usually fitted with stabilizers and in large vessels bow thrusters to assist manoeuvrability at low speeds.

The design of these vessels is highly specialised, the amenities provided being of a high class. The numerous decks are necessary to provide observation and sun decks, verandas, lounges, bars, dining rooms, shops, galleys and bakeries, etc., in addition to accommodation for passengers.

PASSENGER VESSEL

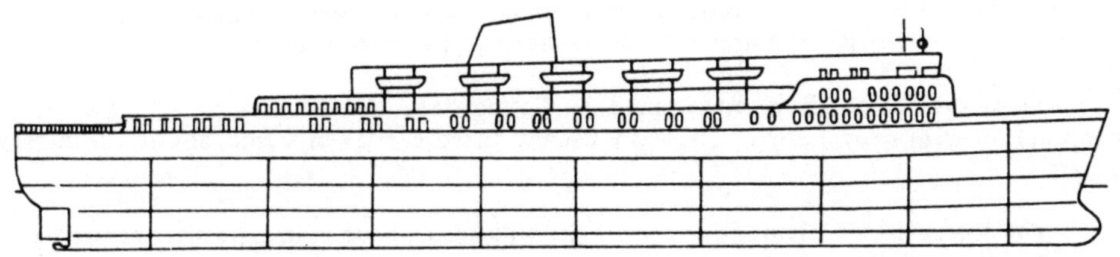

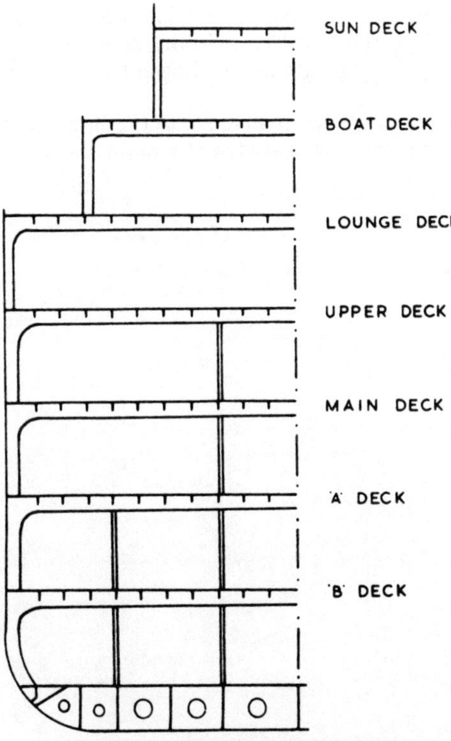

SUN DECK

BOAT DECK

LOUNGE DECK

UPPER DECK

MAIN DECK

'A' DECK

'B' DECK

MIDSHIP SECTION THRO' ACCOMMODATION

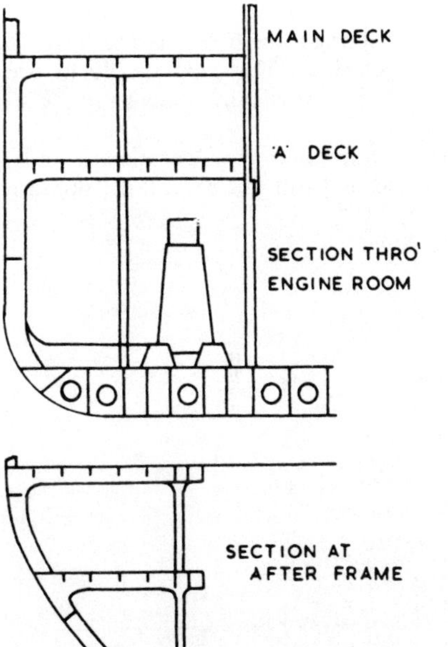

MAIN DECK

'A' DECK

SECTION THRO' ENGINE ROOM

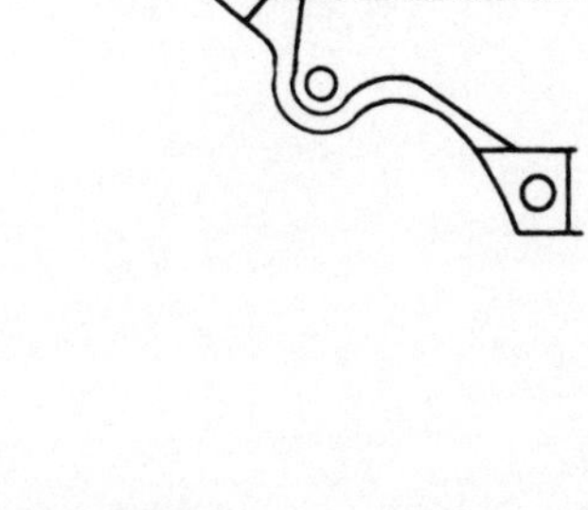

SECTION AT AFTER FRAME

A profile and midship cross section of a roll on/roll off vessel is illustrated. Noticeable are the ramps and doors at the bow and stern, these are detailed on page 87.

An essential feature of this type of vessel is clear decks uninterrupted by transverse bulkheads. Deck heights are to be sufficient to accommodate the various types of vehicles that will be carried. In the type of vessel illustrated the lower decks are used for the carriage of light vehicles, the upper decks for heavy vehicles and trailers.

Transverse strength is maintained by fitting deep, closely spaced web frames in conjunction with deep beams. These are usually fitted every 4th frame, about 3.0 metres apart.

The lower decks, divided by watertight bulkheads, have hydraulically operated and cleated sliding bulkhead doors to facilitate the movement of vehicles.

The deck thickness is increased to take the concentrated loads; a reduction in the spacing of the longitudinals with an increase in size. A centre line row of pillars is fitted.

Ramps are fitted at the bow and stern to facilitate the loading and discharging of the vehicles. The separate decks are reached by fixed or hydraulically operated ramps and lifts. Moveable ramps have the advantage of giving a greater loading area.

Some Ro. Ro. Vessels are built with stern doors set at an angle to the ship's centre line so that full roll on/roll off operations can be maintained alongside a normal quay.

RO-RO FERRY

ACCOMMODATION, LOUNGES ETC.
STERN DOOR AND RAMP
VEHICLE DECK
BOW DOOR AND RAMP
RAMP
ENGINE
ROOM
VEH. DK
CAR DECKS

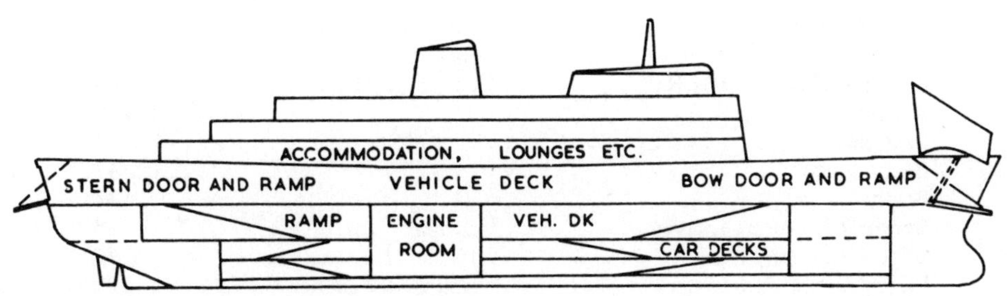

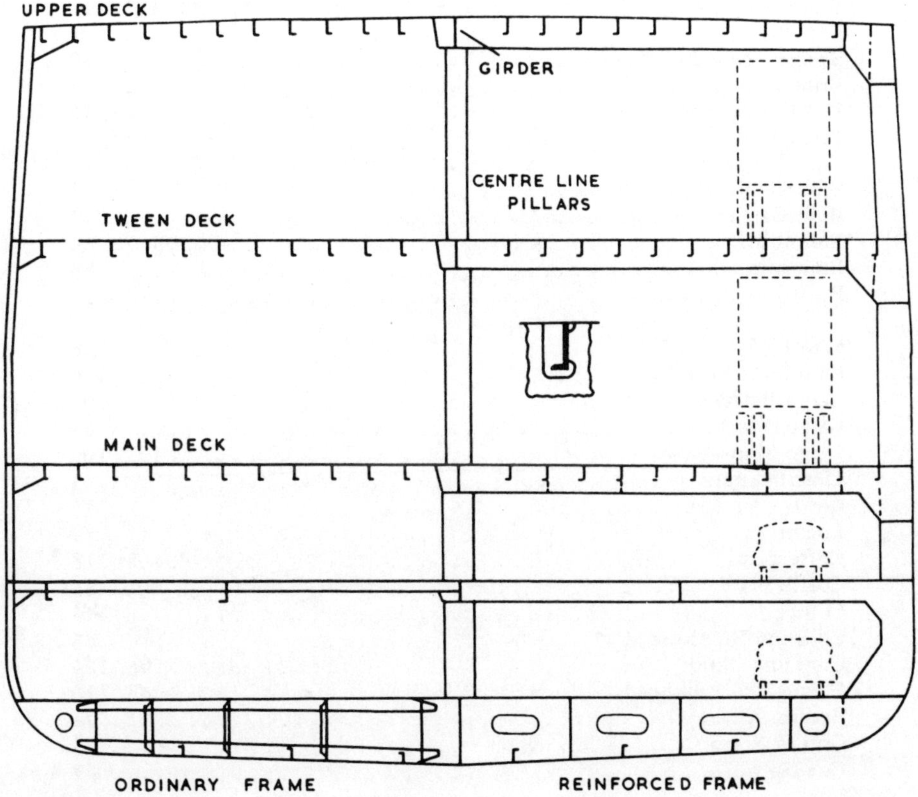

UPPER DECK
GIRDER
CENTRE LINE
PILLARS
TWEEN DECK
MAIN DECK
ORDINARY FRAME
REINFORCED FRAME